The Open University

Science Foundation Course Unit 31

THE NUCLEUS OF THE ATOM

Prepared by the Science Foundation Course Team

THE OPEN UNIVERSITY PRESS

A NOTE ABOUT AUTHORSHIP OF THIS TEXT

This text is one of a series that, together, constitutes *a component part* of the Science Foundation Course. The other components are a series of television and radio programmes, home experiments and a summer school.

The course has been produced by a team, which accepts responsibility for the course as a whole and for each of its components.

THE SCIENCE FOUNDATION COURSE TEAM

M. J. Pentz (Chairman and General Editor)

F. Aprahamian	(Editor)	A. R. Kaye	(Educational Technology)
P. Chapman	(Physics)	J. McCloy	(BBC)
A. Clow	(BBC)	J. E. Pearson	(Editor)
P. A. Crompton	(BBC)	S. P. R. Rose	(Biology)
G. F. Elliott	(Physics)	R. A. Ross	(Chemistry)
G. C. Fletcher	(Physics)	P. J. Smith	(Earth Sciences)
I. G. Gass	(Earth Sciences)	F. R. Stannard	(Physics)
L. J. Haynes	(Chemistry)	J. Stevenson	(BBC)
R. R. Hill	(Chemistry)	N. A. Taylor	(BBC)
R. M. Holmes	(Biology)	M. E. Varley	(Biology)
S. W. Hurry	(Biology)	A. J. Walton	(Physics)
D. A. Johnson	(Chemistry)	B. G. Whatley	(BBC)
A. B. Jolly	(BBC)	R. C. L. Wilson	(Earth Sciences)
R. Jones	(BBC)		

The following people acted as consultants for certain components of the course:

J. D. Beckman	R. J. Knight	J. R. Ravetz
B. S. Cox	D. J. Miller	H. Rose
G. Davies	M. W. Neil	
G. Holister	C. Newey	

The Open University Press
Walton Hall, Bletchley, Bucks

First Published 1971. Reprinted 1972, 1973.
Copyright © 1971 The Open University

Designed by the Media Development Group of the Open University

Printed in Great Britain by
EYRE AND SPOTTISWOODE LIMITED
AT GROSVENOR PRESS PORTSMOUTH

SBN 335 02039 9

Open University courses provide a method of study for independent learners through an integrated teaching system, including textual material, radio and television programmes and short residential courses. This text is one of a series that makes up the correspondence element of the Science Foundation Course.

For general availability of supporting material referred to in this text, please write to the Director of Marketing, The Open University, Walton Hall, Bletchley, Buckinghamshire

Further information on Open University courses may be obtained from The Admissions Office, The Open University, P.O. Box 48, Bletchley, Buckinghamshire.

1.3

Contents

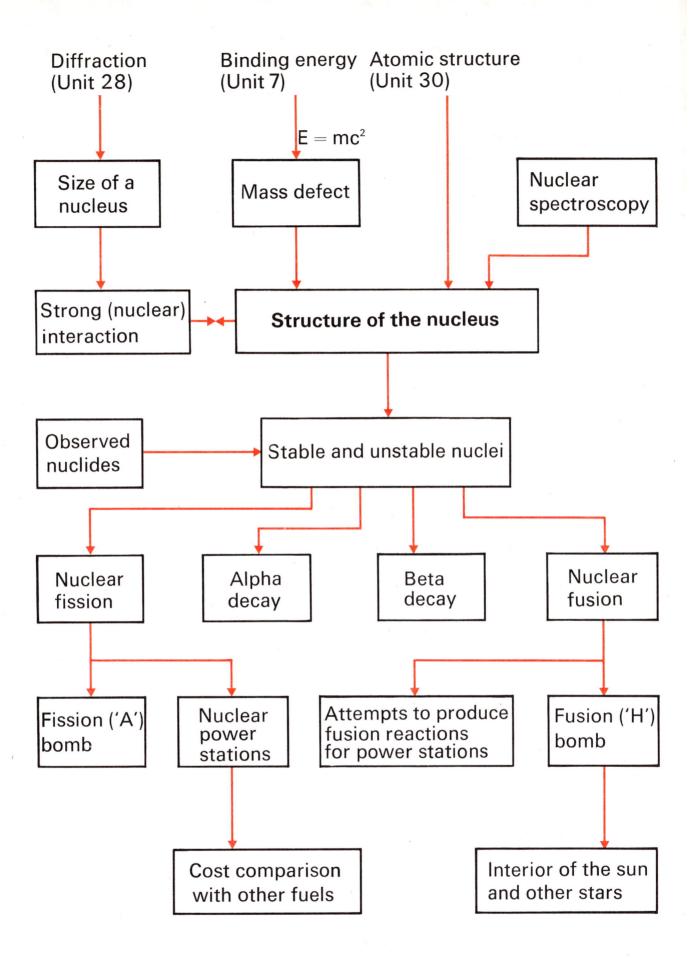

List of Scientific Terms, Concepts and Principles used in Unit 31

Taken as pre-requisites			Introduced in this Unit			
1	**2**		**3**		**4**	
Assumed from general knowledge	**Introduced in a previous Unit**	**Unit No.**	**Developed in this Unit**	**Page No.**	**Developed in a later Unit**	**Unit No.**
diameter	atom	3	effective nucleus radius	11		
	atomic nucleus	6	nucleon	11		
radius	electron wave	29	nucleon density	11		
	momentum	3	strong interaction charge			
volume	wavelength	2	independence	13		
	de Broglie formula	29	nucleon energy level	14		
density	nuclide	6	total nuclear binding energy	17		
	diffraction	2, 28	neutron excess	21		
	interference	28	positron	23		
	femtometre	*HED*	mean binding energy per nucleon	25		
	atomic mass number	6	radioactive series	27		
	proton	6	nuclear fission liquid-drop model	29		
	neutron	6	chain reaction	31		
	relative atomic mass	6	critical mass	31		
	energy	4	control rod	32		
	energy level	6	moderator	32		
	potential difference	4	nuclear fusion	34		
	electron-volt	4				
	ionization energy	6				
	atomic number	6				
	photon	6, 29				
	gamma-radiation	6				
	mass defect					
	rest energy	4				
	radioactivity	2				
	beta-particle	6				
	isotope	6				
	alpha-particle	6				
	half-life	2				

Any scientific terms used in this Unit but not listed above are marked thus † and defined in the glossary

Aims

1 To give an understanding of the atomic nucleus, its size and components, how it is held together, and which kinds of nuclei can exist.

2 To increase understanding about radioactive decay and in particular to link it with nuclear instability.

3 To introduce fission and nuclear fusion and thereby give some understanding of nuclear bombs and nuclear power stations. Also to consider the likely future of nuclear power as an economic proposition.

Objectives

When you have finished this Unit you should be able to:

1 Define correctly or recognize the best definition of each of the terms, concepts and principles in Column 3 of Table A.

2 Recognize true and false statements concerning the range of the 'strong interaction' force and the approximate size of nuclei.

3 Carry out simple calculations relating nuclear radius, atomic number, mean nuclear mass and mean nuclear density.

4 Identify from two angular scattering distributions the larger nucleus or the higher momentum beam; give reasons for the high momentum/energy needed.

5 Give brief evidence for, or recognize evidence for, neutron and proton energy levels in nuclei.

6 Summarize briefly, or recognize valid statements about, the limited range of numbers of protons and neutrons in observed nuclei, both stable and unstable; give reasons for the neutron excess in heavier nuclei.

7 Recognize examples of alpha decay, beta decay, gamma decay, fission and fusion which are either *possible* or *impossible* because of charge and nucleon conservation; predict which of β^+- and β^--decay is likely to occur in a given unstable nuclide from a knowledge of the stable nuclides with the same atomic mass number.

8 Explain from a graph of mean binding energy per nucleon (Fig. 8) why the following processes can take place with a release of energy:

(a) alpha decay or fission of many heavy elements;
(b) fusion of light elements.

9 Calculate the energy (in joules or MeV) released in alpha decay, fission or fusion, given the masses of relevant nuclides (in MeV/c^2 or a.m.u.) and other appropriate constants.

10 Discuss in not more than two hundred words each, or recognize statements about, the relative economic factors of nuclear and conventional power stations, the social or ecological impact of nuclear and conventional power stations, and the likely nature of power stations in the long term.

11 Identify the contexts in which fusion is or could be important, and the problems which at present frustrate fusion power research.

31.1   Introduction.

This Unit will be concentrating on the tiny fragment, the nucleus, which carries most of the mass of an atom. Its very small size and great density can only be explained by a new force—the *strong interaction*. This force has a very short range, even smaller than the diameters of nuclei.

You know by now that many of the properties of the chemical elements are determined by the way their electrons are arranged in energy shells. Many properties of nuclei are accounted for in a similar way, because nuclei can also be regarded as possessing an energy-level structure—in fact two separate sets of levels, one set for protons and one for neutrons. Following on from a discussion of these levels, we can explain why only certain stable nuclides exist. Other nuclides exist with a slightly less restricted range of mass number and charge, but they are all unstable. These give rise to the phenomenon of radioactivity which you met in earlier Units. These forms of instability will be discussed in more detail than before.

The discussion will centre on the ability of the nuclear rest mass to convert into energy, as expressed in Einstein's equation $E = mc^2$. This consequence of relativity theory becomes particularly important in the context of the process of nuclear decay called fission. You will see how one fission process can lead to another in a self-perpetuating chain reaction in which large quantities of energy may be released. This reaction can be controlled, as in a nuclear power station, or it may lead to an explosion as in the 'A-bomb'. Fusion is another nuclear process which gives rise to energy release in particular to the energy of the H-bomb. The difficulties encountered in trying to harness this power for useful purposes are mentioned.

31.2   Nuclear Dimensions

31.2.1   Another step down in size

In Unit 6 you learned how very small atoms are. They are so small that it is barely possible to observe them directly. From the results of many measurements including X-ray diffraction from crystals, we deduce that atomic sizes are around ångström unit, i.e. 10^{-10} m. The atomic nucleus is much smaller still. Its radius is a few times 10^{-15} m, though the exact size varies from one nuclide to another. How can we measure the size of atomic nuclei?

It is impossible to spread a film of bare nuclei on top of a bowl of water (as was discussed for atoms in Unit 6, section 6.12), or to make a crystal with bare nuclei packed tightly together. This is because nuclei are normally surrounded by the envelope of electrons which forms the outer part of an atom.

One technique for measuring nuclear sizes is the scattering of high-momentum particles—particularly electrons.

Why must they be high-momentum particles?

They must be high-momentum particles because, as you saw in Unit 2, the wavelength of the incoming beam (whether of ultra-sound from a bat trying to find a moth, of light in a microscope, or of electrons to observe nuclear sizes) must be smaller than—or comparable to—the size of the scatterer.

How are the momentum p and wavelength λ of any particle-wave related to each other?

The momentum p and the wavelength λ of any particle-wave are related by the de Broglie formula (Unit 29, equation 2):

$$p = \frac{h}{\lambda} \quad \dots\dots\dots\dots\dots \quad (1)$$

If the wavelength λ of the beam is very small, the momentum p must be correspondingly large. In order to 'see' an object 10^{-15} m across, the wavelength of the beam must not be much greater than 10^{-15} m.

Given that Planck's constant $h = 6.6 \times 10^{-34}$ Js, calculate the momentum that such a beam of particles will have.

$$p = \frac{h}{\lambda} = \frac{6.6 \times 10^{-34}}{10^{-15}} = 6.6 \times 10^{-19} \text{ N s}$$

A momentum of the order of 10^{-19} N s may seem very small, but for a tiny particle like an electron it is very large. The electrons discussed in Unit 29, section 29.1 had a wavelength of 10^{-10} m, so their momentum was 10^5 times smaller than that of the high-momentum particles we are considering here.

Very high-momentum particles are produced by modern high-energy accelerators. In Unit 32 you will learn about the CERN proton synchrotron at Geneva, Switzerland. This machine can accelerate protons to a momentum of about 3×10^{-18} N s.

Figure 1

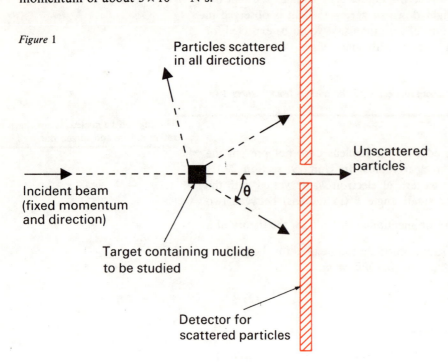

(*a*) *A high-momentum scattering experiment.*

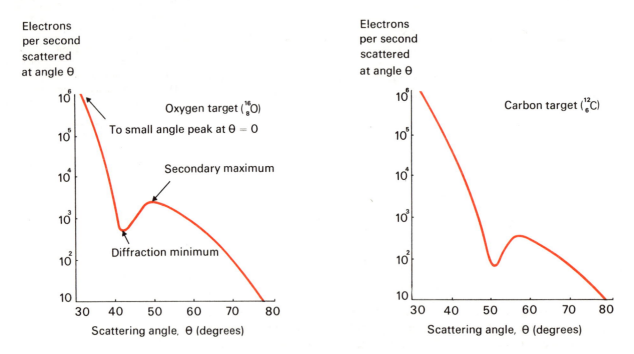

(*b*) *High-momentum electron scattering distributions for* $^{16}_{8}O$ *and* $^{12}_{6}C$.

The size of the nucleus shows up in the angular distribution of the scattered beam. Figure 1 shows the experimental arrangement and results of a typical experiment. You will notice on Figure 1 (b), the increase of the scattering rate at smaller angles (the small angle diffraction peak), the sharp dip at $\theta = 44°$ for oxygen and at $\theta = 51°$ for carbon, and the secondary maximum at larger angles. The shape of the graphs represents an interference pattern in the electron wave, caused by diffraction round the

nucleus. You have seen similar effects in Figure 10 of Unit 28 and in Figure 4 of Unit 29. Figure 5 of Unit 29 shows the angular distribution of light waves diffracted through a slit; the pattern is very similar for an obstacle with sharp edges. However in Figure 5 of Unit 29 the first dip (diffraction minimum) goes right down to zero—no light is observed in this direction. A similar direction of zero intensity would be expected for electrons diffracting round an obstacle with sharp edges.

Can you guess anything about the edges of the nucleus from Figures 1 (b) and 1 (c)?

The edge of the nucleus is not sharp but rather diffuse and spread out.

The lack of sharpness of the edge of the nucleus does not prevent the measurement of an effective nuclear radius. In Unit 29, section 29.1.1 you saw how the interference pattern of electron waves varied with the separation d of two slits; the small angle θ (in radians) between two maxima was given by $\theta = \dfrac{\lambda}{d}$. As we mentioned in Unit 28, the theory of a circular obstacle shows that a similar formula can be used for the angle θ_1 between the *central maximum* and the *first minimum*:

$$\sin \theta_1 = \frac{1.22\lambda}{d} \quad \ldots\ldots\ldots\ldots\ldots \quad (2)$$

where d is the diameter of the obstacle.

Now look at Figures 1 (b) and 1 (c). Both experiments used electrons of the same momentum (2.2×10^{-19} N s) and hence the same wavelength.

Which of the two nuclei is the larger?

The oxygen nucleus. From equation 2, if λ is the same then a larger diameter means a smaller angle of diffraction. Oxygen has a smaller angle for the first minimum (44°) so it is the larger nucleus.

It is easy to calculate the approximate radius of the oxygen nucleus from equations 1 and 2. The wavelength of the electrons is given by equation 1:

$$\lambda = \frac{h}{p} = \frac{6.6 \times 10^{-34}}{2.2 \times 10^{-19}} = 3 \times 10^{-15} \text{ m}$$

Equation 2 now gives the diameter d:

$$\sin \theta_1 = \frac{1.22\lambda}{d}$$

$$\therefore \quad d = \frac{1.22\lambda}{\sin \theta_1} \approx \frac{1.22 \times (3 \times 10^{-15})}{\sin 44°}$$

$$\approx 5 \times 10^{-15} \text{ m}$$

$$\text{and radius } r = 2.5 \times 10^{-15} \text{ m}.$$

This is reasonably close to the value determined by more thorough analysis of the data. All nuclei have radii of this order of magnitude, so the femtometre (1 fm = 10^{-15} m) is a convenient unit of length to use; the radius of the oxygen nucleus is about 2.5 fm.

This kind of analysis has been performed on data from electron scattering experiments with various nuclides in the target. The results can be summed up in one simple formula which is approximately true for the radius r of all but the lightest nuclei:

$$r = 1.2 \, A^{\frac{1}{3}} \text{ fm} \quad \ldots\ldots\ldots\ldots \quad (3)$$

where A is the atomic mass number*.

* *If you are unsure of the meaning of the index in $A^{\frac{1}{3}}$, see* MAFS, *section* 1.A.5.

In Unit 6, section 6.2.2, you learned that A is equal to the sum of the total number of protons plus the total number of neutrons in a nucleus. Both protons and neutrons are called *nucleons*, so A is equal to the total number of nucleons.

nucleons

31.2.2 The nuclear density

The volume V of a sphere is given by

$$V = \frac{4}{3}\pi r^3$$

For approximate calculations $\pi \approx 3$, so

$$V \approx 4r^3 \ldots\ldots\ldots\ldots\ldots \quad (4)$$

What is the volume of a sphere of radius 4 fm?

From equation 4 the volume V is given approximately by:
$$V \approx 4r^3 = 4 \times (4)^3 \text{ fm}^3$$
$$\therefore \quad V \approx 256 \text{ fm}^3.$$

The density of a substance is the mass of a unit volume of that substance. In discussing nuclear densities it is convenient to use the number of nucleons in one cubic femtometre. This can be re-expressed as (mass per fm³) if it is multiplied by the average value of the mass of a nucleon.

Now $V \approx 4r^3$ (equation 4)

and $r = 1.2\, A^{\frac{1}{3}}$ (equation 3)

$$\therefore \quad V \approx 4\,(1.2\, A^{\frac{1}{3}})^3 = 6.9\, A$$

Hence the average density $\rho = \dfrac{A}{V}$ nucleons fm^{-3}

$$= \frac{A}{6.9A}$$

$$= 0.14 \text{ nucleons per cubic femtometre} \ldots\ldots\ldots\ldots (5)$$

How does the average nuclear density depend on the atomic mass number?

It doesn't. The average nuclear density is the same for all nuclei (except the smallest).

The mass of a proton or neutron is about 1.7×10^{-27} kg. Estimate from equation 5 the ratio of the nuclear density and the density of water (1 000 kg m^{-3}).

1 fm³ is $(10^{-15})^3$ m³ or 10^{-45} m³. So the nuclear density expressed in SI units is

$$\rho \approx 0.14 \times \left(\frac{1.7 \times 10^{-27}}{10^{-45}}\right) \text{ kg m}^{-3}$$

$$\approx 0.24 \times 10^{18} \text{ kg m}^{-3}$$

So $\dfrac{\rho \text{ nuclear}}{\rho \text{ water}} \approx \dfrac{0.24 \times 10^{18}}{1000} \approx 2.4 \times 10^{14}.$

This is an enormous number and is a measure of how much ordinary matter could be compressed if all the atoms were squashed into their nuclei. Such a collapse of matter is believed to take place inside certain stars. A five millilitre teaspoonful of such material would have a mass of 1.2×10^{12} kg—about as much as Ben Nevis.

31.2.3 Summary of section 31.2

The size of a nucleus can be measured by observing the diffraction of a beam of electrons round the edges of the nucleus. The electrons must have a high energy and momentum so that their wavelength is small enough to be comparable with, or less than, the nuclear diameter. Experiments lead to radii of a few femtometres (10^{-15} m); for example, the radius of an oxygen nucleus is about 2.5 fm. All nuclei except the lightest have approximately the same nuclear density, so the volume of a nucleus is directly proportional to the number of nucleons (protons and neutrons) in it. This nuclear density is greater than the density of water by a factor of about 2.4×10^{14}. A large mountain, if it could be compressed so that all the nuclei of its atoms were touching, would fit on to a teaspoon.

Now turn to the Self-Assessment Questions and attempt numbers 1 *and* 2 (*p.* 49).

31.3 The Structure of the Nucleus

31.3.1 The strong interaction

In Unit 6 you learned that the nuclear mass is more than 99.98 per cent of the mass of an atom.

How are nuclear masses measured? If you cannot remember, refer to Unit 6, section 6.2.5.

You have seen how the comparatively small size and large mass of the nucleus leads to a very high density, about 2.4×10^{14} times that of water. Can this be accounted for from what we already know about the forces of nature? Gravitation is so weak that the systems it holds together are characteristically hundreds of miles across—or much more. The gravitational attraction between everyday objects, such as people, is negligible, even when they are close together. Yet very massive objects, like the Earth, have a gravitational attraction which is significant many thousands of miles away in space. In nuclei, the protons and neutrons are very close together, but they are also very light, so they exert infinitesimally small gravitational forces. Electrical attraction between protons and electrons maintains atoms which are about 10^5 times as big as nuclei. These forces cannot explain the density of nuclei; a new force has to be inferred, called the *strong interaction*. This force acts between nucleons, and is sufficiently powerful to hold them inside the tiny volume of the nucleus. The action of this force is very closely related to the elementary particles which are the subject of Unit 32. The strong interaction is approximately *charge independent*, i.e. the force is the same for two protons, two neutrons or one of each. Because of detailed quantum theory effects, there are no nuclides with just two neutrons bound together—but there is a nuclide with one neutron and one proton. This is called deuterium ^2_1H (or ^2_1D). If a proton could be added to this it would stick, forming a helium isotope ^3_2He. If instead, a neutron were added to ^2_1H, it also would stick, forming ^3_1H—tritium, another isotope of hydrogen.

strong interaction

charge independence

What practical problem can you anticipate in trying to add a proton to ^2_1H? Why would it be easier to add a neutron?

The strong interaction or 'nuclear force' was first mentioned in Unit 4, section 4.2.3. It was argued there that it must be a short-range force, if it could influence objects some distance away from the immediate region of the nucleus, it would swamp the effects of the gravitational and electromagnetic forces. This short range is consistent with the small size of the nucleus. It also helps to explain why the density of large nuclei is independent of the mass number A. In Figure 2 you can see how a nucleon deep inside a large nucleus is acted upon by the short-range force (range $\approx$ 2 fm) of its nearest neighbours. It doesn't matter much what kind of nucleus the nucleon is in. The protons and neutrons always have similar properties, so the range and strength of the force are the same. The nucleons therefore pack themselves together with about the same density

To add a proton to a nucleus it is necessary to overcome the long-range electrical repulsion between the positive charge of the proton and the positive charge of the nucleus. A neutron has no charge, so it experiences no electrical repulsion as it approaches a nucleus.

in nuclei of differing sizes. Only in the surface regions is there any variation in the forces on individual nucleons.

Figure 2 shows, schematically, the strong interaction forces acting on two nucleons A and B. Notice that the forces on A balance, but B is pulled inwards. Also A has links with three other nucleons, while B has links with only two.

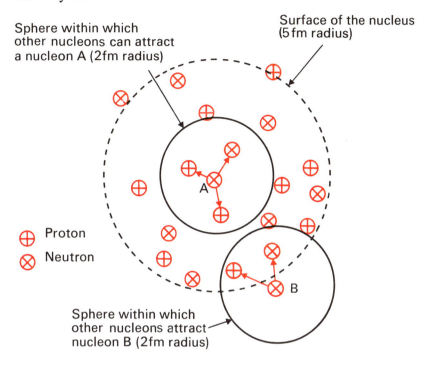

Figure 2   *The strong interaction forces on different nucleons.*

31.3.2   Packing neutrons and protons together

In Units 6, 7 and 30, you have become familiar with the energy levels that are available to electrons when they are packed together in an atom. There is strong evidence that nucleons also have a structure of energy levels within the nucleus. However this structure is created by the strong interaction which is not yet fully understood.

> **What is the difference between the binding energy of a nucleon and the ionization energy of an atomic electron?**

The two quantities have essentially the same physical significance, but very different typical values. The ionization energy of an atomic electron will be in the region of 10^{-19} joule: 22×10^{-19} J for the ground state of hydrogen for example. This means that the binding energy of this electron —the energy the atom would need to be turned into an H$^+$ ion—is 22×10^{-19} J.* Binding energies are often expressed in units called electron-volts; one electron-volt (1eV) is the energy an electron (or proton) acquires when accelerated through a potential difference of one volt, and

---

* *The ionization energy of an atomic electron is the amount of energy it has to be given to remove it from the atom. Another way of saying the same thing is that the atomic electron is bound to the atom with a 'binding energy' equal (numerically) to its ionization energy.*

An analogy might perhaps help you to understand this idea. Suppose you are put in prison because you can't pay bail of £5. It needs £5 to bail you out. You are kept in by a lack of £5. That £5 you haven't got is your 'binding money', and if someone supplies it and you get let out, then the £5 he supplies is your 'ionization money'.

is equal to 1.6×10^{-19} J (see Unit 4, section 4.4.3). The binding energy of the electron in a hydrogen atom in its ground state is 13.6eV. The binding energy of a nucleon in a nucleus is roughly a million times bigger than this; 6.75MeV for the first neutron to be removed from $^{21}_{10}$Ne, for example.

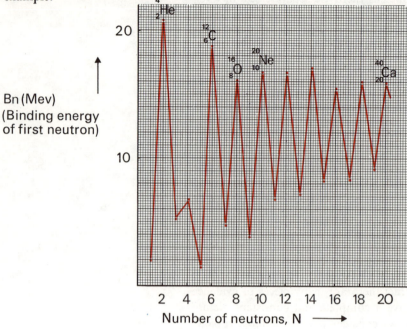

Figure 3   *The binding energy of the first neutron plotted against the number of neutrons* N.

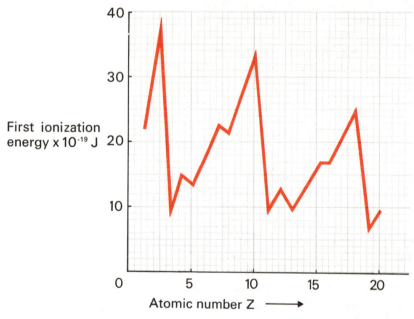

Figure 4   *The ionization energy (of the first electron) plotted against atomic number.*

Figure 3 is a plot which looks very similar to Figures 4 and 5 of Unit 7. (Figure 5 of Unit 7 is repeated here as Figure 4.) Figure 3 is a plot of the energy which would be required to remove the first neutron from each of a whole series of nuclei (the binding energy). This is plotted against the number of neutrons, *N*, in each nucleus.

What are the most obvious similarities between Figures 3 and 4?

Both of these plots are very jagged and the peaks are always at even numbers. Clearly, some kind of energy-level structure is indicated for neutrons.

15

Proton binding energies vary with the number of protons Z in a very similar way to the variation of neutron binding energies with N in Figure 3. So protons appear to have an energy-level structure too. The theory of these levels is worked out in a similar manner to the theory of atomic electron energy levels, but in the nucleus there are separate sets of levels for neutrons and for protons. Just as atoms with many electrons have a number of fully occupied electron levels, so large nuclei with many nucleons have a number of fully occupied nuclear levels.

> **It has been stated that the strong interaction is independent of charge. What does this imply about the relationship between proton and neutron energy levels?**

Since the major force binding the nucleons is independent of charge, the proton and neutron energy levels can be expected to be very similar to one another (an important difference will be discussed in section 31.4.1).

31.3.3 Nuclear spectroscopy

So far, Figure 3 is the only evidence we have given to support the idea of energy levels in the nucleus. In fact there are many pieces of evidence which can be pieced together to give a complete and consistent picture of the energy levels. In this section we consider one of the most copious sources of this evidence.

When an *atom* is *excited* by a collision or by absorption of a photon, an electron changes from a state of lower energy to one of higher energy. This leaves a vacancy in a lower level until one or more electrons change state downwards (in energy) to fill the vacancy and leave the atom in its ground state again. Each of these changes of state is accompanied by the radiation of a photon of light.

> **Can you think of the nuclear parallel to this process?**

Nuclei may become excited in a number of ways—by collision, by absorption of a photon, as a result of beta-decay and by other mechanisms. One way in which an excited nucleus can return to its ground state of lowest energy is by radiating the excess energy in the form of photons. These photons have higher energy than light or X-rays and are called *γ-rays*. (The energies of the photons in visible light are in the region of 1eV; X-ray photons have energies in the region of 1 KeV; γ-ray photons have energies from about 0.2 MeV upward.) Just as the major sources of information about atomic energy levels are optical, ultra-violet and infra-red spectra, so γ-ray spectroscopy yields vast quantities of data on nuclear energy levels.

γ-rays

Figure 5 shows some of the observed energies of γ-rays emitted from excited $^{31}_{15}$P. The horizontal lines represent energy levels in MeV above the ground state, and the length of each vertical line gives the energy of a particular γ-ray spectral line. Compare Figure 5 with the diagrams (in Unit 6, section 6.5.2 and Unit 7, 7.2) of energy levels in many-electron atoms. In Unit 6, you learned that when an electron 'jumps' from one energy level to another inside an excited atom, a photon is emitted. Similarly in a nucleus, when a nucleon 'jumps' from one energy level to another a γ-ray photon is emitted. This diagram is very similar to Figure 7 in Unit 6, though the photon energies here are a million times larger than the energies of photons emitted by the hydrogen atom.

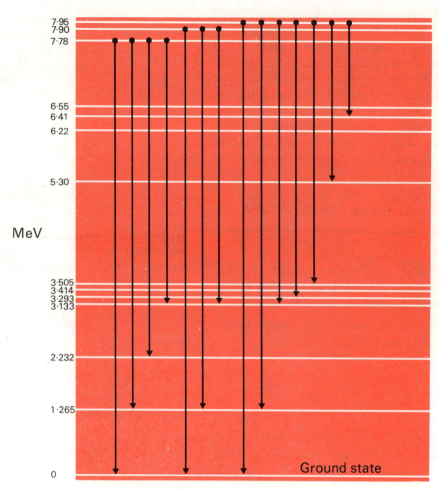

MeV

Figure 5   *Some energy levels and transitions in* $^{31}_{15}$P.

31.3.4   Unpacking the nucleus—binding energy and mass defect

How much energy would be required to pull a nucleus apart, one nucleon at a time?

A certain amount of energy will be needed to remove each nucleon, in turn, from the nucleus. The sum of all these energies is the amount of energy required to take the nucleus completely apart. This is the *total binding energy* of the nucleus—the energy required to remove all of its nucleons to a large distance from one another—out of range of the strong interaction and to a separation where electrostatic interactions are negligible. The total binding energy of a nucleus is a concept you met briefly in Unit 6. It gives rise to the *mass defect* which can be measured with a mass spectrometer. Thus, bearing in mind that mass and energy are equivalent quantities (Unit 4, rest mass energy $= M_0c^2$), the binding energy of the nucleus is that amount of energy which is equivalent to the mass defect of the nucleus, using the relativistic formula $E = Mc^2$.

total binding energy

mass defect

You will remember that in Unit 6, section 6.2.6, the mass defect of a nucleus was defined as the difference between its mass and the masses of its constituent nucleons. When a structure is bound together, energy is needed to pull it apart and this energy is equal in magnitude to the total binding energy. So the total binding energy is equal to the reduction in mass of the bound system. In studying nuclear physics, there are many processes in which part of the mass of some object becomes available in the form of kinetic energy, or in which energy is absorbed, so increasing the mass of an object. It is therefore convenient to work in a system of units in which the mass-energy equivalence is easily expressed.

To use the SI units of joules and kilogrammes would be very clumsy, requiring repeated conversions between them and the manipulation of large powers of ten. We have already suggested above (section 31.3.2) that nuclear energies are conveniently measured in millions of electron-volts (MeV). Masses are proportional to energies ($E = mc^2$), so the same basic unit can be used; the mass unit is then MeV/c^2. This means that the mass is expressed in terms of the energy it would produce if totally converted. A mass of one MeV/c^2 means the mass which could produce 1 MeV of energy if totally converted; if a value in kg is ever required, then the MeV is converted to joules and divided by the square of the velocity of light.

The mass of the $^4_2$He nucleus is approximately 3 750 MeV/c^2. Calculate the equivalent mass in kg.

Table 1 gives the masses of some particles and nuclei expressed in units of MeV/c^2, in atomic mass units, and in kg. The table is given mainly as data for the *SAQ*s. You are not expected to remember any of the numbers.

One advantage of using MeV/c^2 is that the total binding energy of a nucleus (in MeV) is numerically equal to the mass defect (in MeV/c^2).

31.3.5 Which nuclei exist?

Figure 6 (a) shows the values of Z (number of protons), N (number of neutrons) and A (number of nucleons) for all the nuclides with Z less than 31 which have ever been observed. (You will remember from Unit 6 that each element has its own characteristic value of Z, but N varies from isotope to isotope of the element.)

$1 eV = 1.6 \times 10^{-19}$ J (approximately), and 1 MeV $= 1.6 \times 10^{-13}$ J

The velocity of light $c = 3 \times 10^8$ m sec^{-1} (approximately)

So the mass of ^{4}He

$$= \frac{3750 \times 1.6 \times 10^{-13}}{(3 \times 10^8)^2}$$

$$= \frac{3.75 \times 1.6}{9} \times 10^{-26}$$

$$= 6.67 \times 10^{-27} \text{ kg}$$

(approximately—see Table I for more exact values.)

Figure 6

(a) Known nuclides with Z $\leqslant$ 30.

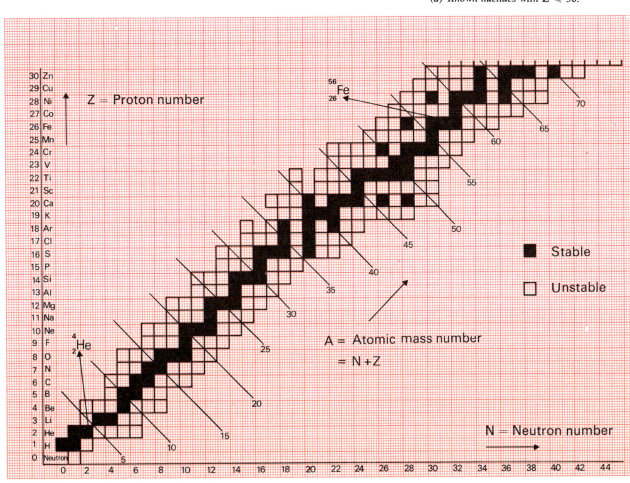

Table 1 Masses of some Particles and Nuclides*

Symbol	Name	Stable or unstable	Mass in MeV/c²	Mass in atomic mass units (a.m.u.)	Mass in kilogrammes	Binding energy per nucleon of stable nuclides	Comment
e	electron	stable	0.51106	0.000549	9.109×10^{-31}	—	
p	proton	stable	938.256	1.0073	1.6726×10^{-27}	0	
n	neutron	unstable	939.550	1.0087	1.6748×10^{-27}	(0)	decays $n \rightarrow p + e^- + \nu$
^1_1H	hydrogen	stable	938.767	1.0078	1.6734×10^{-27}	0	$p + e^-$
^2_1H	deuterium (heavy hydrogen)	stable	1876.1	2.0014	3.3231×10^{-27}	1.1	least tightly bound stable nuclide
^4_2He	helium	stable	3749.1	4.0026	6.6459×10^{-27}	7.1 MeV	α-particle + 2e$^-$
^6_2He	helium	unstable	5607.5	6.0189	9.9938×10^{-27}	5.2 MeV	Same A value
^6_3Li	lithium	stable	5604.0	6.0151	9.9875×10^{-27}		Same A value
^8_4Be	beryllium	unstable	7457.0	8.0053	1.3292×10^{-26}		decays to 2 α-particles
$^{12}_{6}C$	carbon	stable	11178.0	12.000	1.9925×10^{-26}	7.6 MeV	12.0 *a.m.u. by definition*
$^{16}_{7}N$	nitrogen	unstable	14910.0	16.0038	2.6630×10^{-26}	8.0 MeV	Same A value
$^{16}_{8}O$	oxygen	stable	14899.0	15.9949	2.6558×10^{-26}		Same A value
$^{56}_{26}Fe$	iron	stable	52102.0	55.9249	9.2875×10^{-26}	8.8 MeV	One of the most tightly bound of all the nuclides
$^{232}_{90}Th$	thorium	unstable	2.1607×10^5	231.9618	3.8515×10^{-25}		α-emitter
$^{235}_{92}U$	uranium	unstable	2.1886×10^5	234.9561	3.9012×10^{-25}		α-emitter and fissionable
$^{238}_{92}U$	uranium	unstable	2.2164×10^5	237.9492	3.9509×10^{-25}	(7.6 MeV)	α-emitter and fissionable

* *This data is taken from a number of sources. There is some variation on the least significant figures, especially for unstable nuclides. Note that for complete atoms the mass given is that of the neutral atom including its electrons. The unit of mass, MeV/c², is explained in section 31.3.4.*

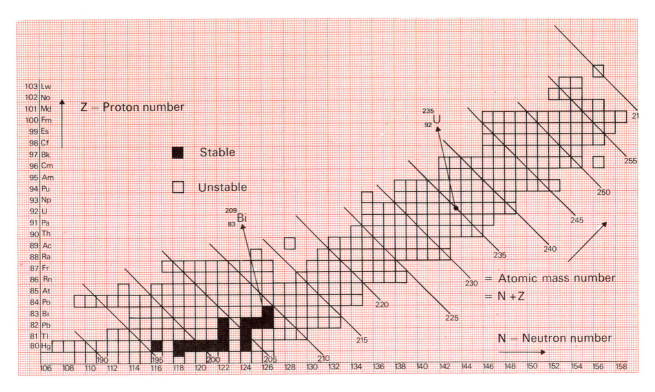

Figure 6 (b) Known nuclides with Z ⩾ 80.

Figure 6 (b) shows the nuclides which have been observed with Z greater than 79. There are, of course, many other nuclides between $Z=30$ and $Z=80$. All we want you to notice at the moment is that some of the nuclides in Figures 6 (a) and 6 (b) are stable (shown shaded), but many are unstable (shown unshaded) and decay (by alpha-(α) or beta-(β) radiation). Most of the nuclei in Figure 6 (b) (high Z) are unstable.

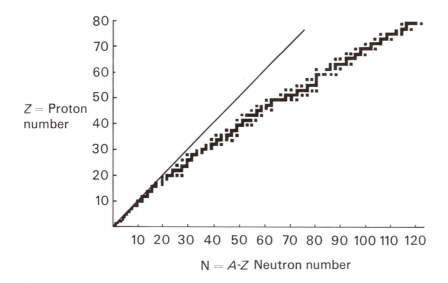

Figure 6 (c) Plot of stable nuclides with Z ⩽ 80.

The vertical axis of each chart gives the number Z of protons in the nucleus. The horizontal axis gives the number of neutrons, $N=A-Z$. Figures 6 (a) and 6 (b) are complicated and incomplete, so it is difficult to see the broad outlines of the picture. Figure 6 (c) is a simpler plot with stable nuclides only up to $Z=80$. Notice that they form a narrow band. Each nuclide is plotted as a small square against the proton number (Z) and

20

the neutron number (N). The full line is at 45° to each axis.

> **For a given Z, what is the stable value of N likely to be? Does the ratio of Z to N for stable nuclei vary with position in the band? Does the observed stable band go on and on as Z increases?**

For a given Z, up to around 20, N is roughly equal to Z in the stable band. As Z increases above 20, N increases somewhat faster than Z. Apparently, heavy nuclei need more neutrons for stability. Thus for

$$^{16}_{8}\text{O (oxygen)}, \quad \frac{Z}{N} = \frac{8}{8} = 1.0$$

$$\text{and for } ^{40}_{20}\text{Ca (calcium)} \quad \frac{Z}{N} = \frac{20}{20} = 1.0$$

$$\text{but for } ^{56}_{26}\text{Fe (iron)}, \quad \frac{Z}{N} = \frac{26}{30} = 0.866$$

$$\text{for } ^{107}_{47}\text{Ag (silver)}, \quad \frac{Z}{N} = \frac{47}{60} = 0.783$$

$$\text{and for } ^{208}_{82}\text{Pb (lead)}, \quad \frac{Z}{N} = \frac{82}{126} = 0.65$$

(Each of the above examples is the most abundant isotope of the element mentioned.)

For the heavier stable nuclei, we say there is a *neutron excess* of $N-Z$. Finally, at $Z=83$ (bismuth) we find the last stable nucleus. At higher values of Z there are some nuclides such as uranium and radium which occur naturally, but all of them are radioactive. They decay with time and (as you saw in Unit 2, section 2.5.4, and in Unit 23, section 23.5.2) they give us important techniques for measuring the age of the Earth and the solar system. Above $Z=105$, no further nuclides have been found or made, though a few more very short-lived examples may yet be produced in the laboratory.* So we find that only a very limited number of nuclei exist. Nucleons cannot be added together indefinitely to make extremely large nuclei, and even with moderately-sized nuclei there are strict limitations on the permitted range of values for the ratio of the number of neutrons to protons. This at first seems rather strange. After all, we have so far thought of the nucleus as consisting of two almost independent systems— an energy-level structure for neutrons and another for protons. How then can the number of protons in the proton energy-level structure affect the number of neutrons to be found at each neutron energy level? Why not have a nucleus with 50 neutrons and no protons, for example? The next section is concerned with explaining why the neutron to proton ratio has only limited values; the two subsequent sections deal with the problem of why there are no extremely large nuclei.

neutron excess

31.3.6 Summary of section 31.3

The nucleus of an atom is held together by a force between nucleons called the 'strong interaction'. Its range is very short (about 2 fm) and it is charge independent—i.e. it does not distinguish between protons and neutrons. Within this short range the strong interaction completely dominates electrical and gravitational forces. The neutrons inside a nucleus are

---

* *Note added in March 1971. A recent experiment claims to have observed a nuclide with $Z=112$.*

distributed among a series of energy levels analogous to the energy levels of electrons in an atom; protons have a similar but separate set of energy levels.

The evidence for nuclear energy levels comes from nuclear spectroscopy. A nucleon is raised to a higher energy level; when it returns to its original stable level it emits a photon of gamma radiation having an energy corresponding to the difference between the two energy levels. The process is analogous to the production of spectral lines of atomic spectra, but the energies involved are much larger.

The nuclear binding energy is equal to the energy which would have to be put back into the nucleus in order to pull it apart into separate nucleons. This binding energy shows itself as a mass defect; the mass of the nucleus is less than that of its constituent nucleons. The binding energy and the mass defect are related by Einstein's equation.

Observed nuclides have a limited range of values of the number of protons (Z) and the number of neutrons (N). For nuclides with Z up to about 20, N is roughly equal to Z. Larger nuclides have a neutron excess. The largest stable nuclide is bismuth with $Z=83$. Above this, all nuclides are radioactive. Above $Z=112$ no nuclides have yet been observed.

Now turn to the Self-Assessment Questions and attempt numbers 3, 10 *and* 11 (*pp.* 49, 52).

If you wish to follow up topics mentioned in this section, the following two books might be useful.

T. Littlefield and N. Thorley, *Atomic and Nuclear Physics*, Van Nostrand, 1968.
J. Fremlin, *Applications of Nuclear Physics*, English U.P., 1964.

31.4 Radioactive Decay

31.4.1 Beta decay—adjusting the ratio of N to Z

We begin by considering nuclear binding energies. Because nuclear binding energies are so large, slight differences in the energy levels of protons and neutrons show up as measurable differences in the atomic mass.*

> **How is the mass of a nucleus different from the mass of the corresponding nuclide?**

The mass of the nucleus will be equal to the atomic mass *minus Z times the electron* mass.**

> **Write down an expression for the energy equivalent of the total rest-mass of a nucleus containing Z protons and N neutrons, in terms of the mass of a proton, the mass of a neutron, and the mass defect.**

The total rest energy is the total rest mass of a nucleus $\times c^2$. This is equal to (the mass of Z protons plus the mass of N neutrons minus the mass defect) $\times c^2$. In Figure 7, we show the value of the total rest energy of the nucleus for a series of nuclides that have $A = 101$.

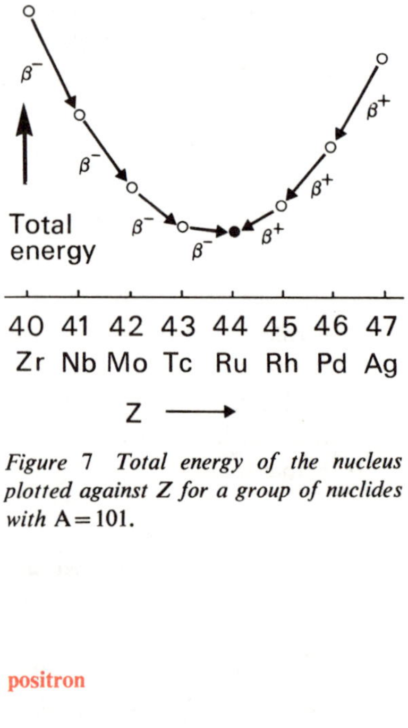

Notice that the variable along the bottom of the graph is Z, so that each of the values plotted is for a different element. The chemical symbol for the element is included in each case below its Z value.

In Figure 7 there are arrows labelled β^+ and β^- which lead down the slope on either side and end at the black dot of $^{101}_{44}$Ru, which is the lowest of all the dots. You met β^- in Unit 6. When a nucleus undergoes β^--decay it emits an electron. It also emits a neutral particle called a neutrino (denoted by the Greek letter Nu, ν). The neutrino is not a nucleon and its loss does not change Z, N or A. We shall more or less ignore it for the moment, though it will always be written in a decay formula. For instance β^--decay in any nucleus can always be regarded as the decay of a bound neutron (denoted by n) into a proton (denoted by p) an electron (denoted by e$^-$) and a neutrino. This decay is written:

$$n \rightarrow p + e^- + \nu.$$

Similarly the β^--decay of tritium $^3_1$H is written:

$$^3_1\text{H} \rightarrow {}^3_2\text{He} + e^- + \nu.$$

Some nuclei decay by β^+-decay. This is the emission of a positive electron— a *positron* (denoted by e$^+$). It is a particle having the opposite charge but

Figure 7 Total energy of the nucleus plotted against Z for a group of nuclides with $A = 101$.

positron

---

* For instance, $^{12}_6C$ has a total binding energy of about 91 MeV. A difference of as little as one per cent in this amount of energy (due to a slight 'excitation' of the nucleus, in which one or more nucleons are not in their most tightly bound, lowest energy levels) would be 0.91 MeV. This is nearly double the rest energy of an electron (0.51 MeV). It would show up as an increase of the mass of the nucleus by an amount nearly equal to the rest mass of two electrons. This would certainly be measurable.

** Electron binding energies are very small compared with the electron rest energy, so we can neglect them.

the same mass as a negative electron. The basic reaction in β^+-decay is:

$$p \rightarrow n + e^+ + \nu.$$

What would you expect to be the effect of β^+ emission on the mass number and atomic number of the nucleus?

The mass number would remain the same but the nuclear charge would decrease by one unit; i.e. the atomic number Z would decrease by one unit. β^+-decay does not happen in naturally occurring nuclides, but it is common among the short-lived isotopes produced by modern techniques. Thus, when an artificially produced atom of $^{101}_{47}$Ag decays by β^+-emissions, it forms a different nucleus.

Which element? Which isotope?

From Figure 7 you can see that the answer is $^{101}_{46}$Pd.

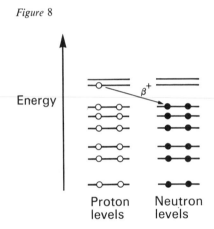

Figure 8

You now have the evidence. What do you think is the explanation of the absence of nuclides away from the stable band of Figure 6 (c)?

The reason is that, although they can be formed, they decay so quickly that they are rarely found in nature.

In the collision of a beam of particles from an accelerator, or in the centre of a star, a group of Z protons and N neutrons may come close together and stick to form a nucleus. For this total number of nucleons, i.e. this value of A ($=Z+N$), there will be just one region of values of Z where the total rest energy of the nucleus is close to a minimum, giving a stable nucleus. The chances are that the group of nucleons will not at first have the best value of Z for stability. If Z is too big, the nucleus has too many protons. A β^+-decay, or a series of them, will correct this. β^+-decay occurs when a proton in the highest occupied energy level of the proton structure changes to a neutron (see Fig. 8 (a)).

This neutron takes the lowest vacant place in the neutron energy-level structure. The process can only happen if the energy level for the newly formed neutron is somewhat lower than the level formerly occupied by the proton that decayed (otherwise energy would have been created out of nothing, which is impossible). If Z on the other hand is too small, the nucleus has too many neutrons and β^--decay will bring it back to the stable region.

In β^--decay, what kind of transition is made?

β^--decay occurs when a neutron in the highest occupied neutron level can drop to a lower energy by becoming a proton (see Fig. 8 (b)).

You may be worried that the total energies of different nuclides shown in Figure 7 are not directly comparable because of the rest energy of the β-particle emitted. Detailed calculations include this and the values of the figure have been corrected to allow for it.

What relation do you notice between the value of Z at the minimum in Figure 7 and the value of the atomic mass number $A=101$?

o Protons

● Neutrons

(a) A simplified picture of the nuclear energy levels in β^+-decay.

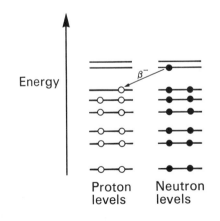

o Protons

● Neutrons

(b) A simplified picture of the nuclear energy levels in β^--decay.

At this particular value of A, the most stable value for Z corresponding to the minimum total energy of the nucleus is 44, that is somewhat less than $A/2$. This is the effect you saw in Figure 6 (c). The stable band swings away from the line $Z=N$, giving a neutron excess in heavy nuclei. In light nuclei stability occurs when $Z=N=\frac{1}{2}A$ (approximately). Apparently neutrons and protons have a very similar energy-level structure when Z is less than 20. But when Z increases beyond 20, the electrical repulsion between the protons has an ever-increasing effect. The energy levels that are available to protons are raised above the equivalent levels for neutrons.* Thus, even when there are more neutrons than protons in the nucleus (as in $^{101}_{47}$Ag, $^{101}_{46}$Pd and $^{101}_{45}$Rh in Figure 7), the highest occupied proton level can be higher than the available neutron levels, and β^+-decay can take place.

Now we have explained why the band of stable nuclei in Figure 6 is quite narrow and why it veers off towards a neutron excess at higher values of Z and N. But why does the band stop? We must next explain why there are no stable nuclides above $^{209}_{83}$ Bi, and no known unstable ones above $Z=112$.

31.4.2   Alpha-particle decay

Look at Figure 9 and try to work out an explanation for the upper limit on nuclear masses from the behaviour of the mean binding energy per nucleon.

You will notice that the *mean binding energy per nucleon* reaches a maximum around $A=65$ and then falls off towards $A=240$. The mean binding energy per nucleon is just the total binding energy of the nucleus divided by the number of nucleons A. Notice how prominent the nuclei $^4_2$He, $^{16}_8$O and $^{12}_6$C are. These nuclei are particularly stable.

mean binding energy per nucleon

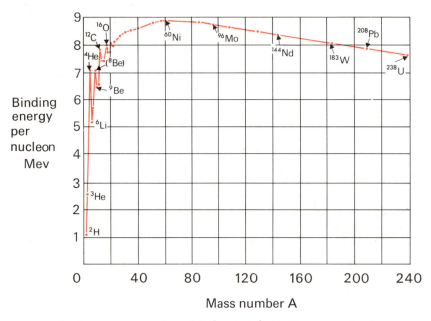

Figure 9   Binding energy per nucleon plotted against the atomic mass number A.

In the region of mass numbers above $A=56$, two new kinds of instability become important. (They are really two variants of the same process but

---

* If this puzzles you at all, think of it like this; remember that the more tightly bound the nucleon the lower is the energy level it occupies. The electrostatic repulsion between the protons makes them less tightly bound to each other and to the nucleus as a whole. So they occupy higher energy levels than the neutrons do.

it is usual to treat them as distinct. The second kind will be discussed in section 5.1.)

Experiments have shown that when $^{238}_{92}$U decays to $^{234}_{90}$Th, 4.2 MeV of energy is released:

$$^{238}_{92}\text{U} \rightarrow \,^{234}_{90}\text{Th} + \,^{4}_{2}\text{He} + 4.2 \text{ MeV}$$

This is the first of the two important kinds of instability at high masses and is called α-*decay* because it involves the emission of an α-particle, or helium nucleus, $^{4}_{2}$He. The energy is liberated because the mean binding energy per nucleon in the two lighter nuclei, $^{234}_{90}$Th and $^{4}_{2}$He, is greater than that in the heavier nucleus, $^{238}_{92}$U, that is, the nucleons are more strongly bound in the two lighter nuclei than they are in the heavier one. In other words, the 92 protons and the 146 neutrons in the $^{238}_{92}$U nucleus have a greater total energy than the 90 protons and 144 neutrons in $^{234}_{90}$Th and the 2 protons and 2 neutrons in $^{4}_{2}$He. The difference is precisely 4.2 MeV. α-decay also happens to some unstable nuclei with low mass. For instance, $^{8}_{4}$Be splits rapidly into two α-particles:

$$^{8}_{4}\text{Be} \rightarrow \,^{4}_{2}\text{He} + \,^{4}_{2}\text{He} + 0.09 \text{ MeV}$$

You may just about be able to see from Figure 9 that the mean binding energy per nucleon of $^{4}_{2}$He is a little greater than the mean binding energy per nucleon of $^{8}_{4}$Be.

The reason is that in $^{4}_{2}$He all the nucleons are in the lowest possible energy level, which is full. In $^{8}_{4}$Be, four of them have to go into the next level, which is much less tightly bound.*

On Figures 6 and 10 we can represent β+-, β−- and α-decay as arrows which move across the Z-N plot. A β+-decay goes in this ↘ direction. A β−-decay goes in this ↖ direction. And an α-decay goes in this ↙ direction.

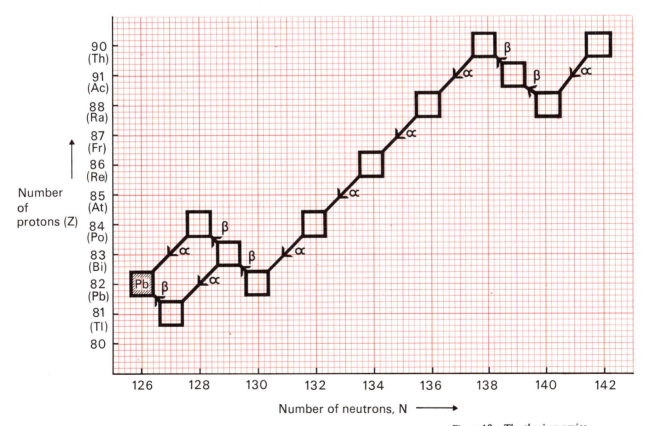

Figure 10   *The thorium series.*

* *Remember, once again, that the lower the energy level occupied by a nucleon, the greater is its binding energy. The four neutrons and four protons in $^{8}_{4}$Be between them have more energy than the four neutrons and four protons in the two helium nuclei ($^{4}_{2}$He + $^{4}_{2}$He). Therefore the binding energy per nucleon in $^{8}_{4}$Be is less than the binding energy per nucleon in $^{4}_{2}$He. So $^{8}_{4}$Be is unstable, and splits into two stable α-particles.*

In Figure 10, you see that happens to a whole family of unstable nuclides called the 'thorium series'. This is one of four families to which all of the nuclei above $Z=82$ belong. None of the α, β^+ and β^- transitions within one of these decay series will make a *daughter nucleus** in one of the other series.

daughter nucleus

Why do you think this is?

β^+- and β^--decays do not change the value of the atomic mass number, A, and α-decay reduces A by 4 units. All the members of a particular decay series will therefore be connected by steps of zero or 4 in A. If the series starts at 238, then it will include nuclides with $A=234$, 230, 226, 222, etc.

Write down what you think the A values of members between 220 and 240 of the other three series will be.

The other series have 220, 224, 228, 232, 236, 240, or 221, 225, 229, 233, 237, or 223, 227, 231, 235, 239. These decay series are important in geology, since they occur in many minerals. The half-lives of α-decay processes, in particular, are often measured in thousands or even millions of years. Other decays, especially β^--decays, may go very rapidly. In a particular mineral sample, it is frequently possible to find a significant amount of the parent nuclide of a series. The daughter nuclides all the way down to the stable end-product—usually a lead isotope—will also be present in quantities depending on their rate of production and decay.

What else might affect the abundance of daughter nuclides?

Some of the daughters will be gaseous—certainly the $^4_2$He formed from α-particles. In certain series there are also isotopes of a gas called radon or radium-emanation with $Z=86$. This word 'emanation' gives the answer to the last question. Even when the mineral is solid, gaseous atoms may diffuse away from their point of production and escape. Thus, in a mineral sample, the abundance of daughters below radon in the series will depend on the ease with which radon has been able to escape from the sample. If you would like to know more about how α-decay takes place, turn to Appendix 1 (Black).

How are airships and North Sea gas connected with the properties of radioactive decays?

An answer to this question can be found in Appendix 3 (Black).

31.4.3   Summary of section 31.4

The total rest energy of a nucleus is measured by its mass. For a series of nuclides having the same value of atomic mass number A, the total rest energy is a minimum for some value of atomic number Z. This minimum determines the value(s) of Z for which the nucleus is stable. For light elements it occurs when Z is approximately equal to $\frac{1}{2}A$.

A nucleus which is not stable may decay to a stable nucleus by changing its value of Z in one of two ways. First, if Z is too small β^--decay occurs. A neutron changes to a proton and a (negative) electron which is ejected from the nucleus. The nucleon moves to a vacant energy level in the

* A daughter nucleus is the nucleus that is left when a decay has taken place.

nucleus which is lower than its original energy before the decay. If, on the other hand, Z is too large, a proton changes to a neutron, and positron is ejected. This is β^+-decay.

Secondly, heavy elements such as uranium may decay by the emission of a helium nucleus (called an alpha-particle). Alpha-decays have much longer half-lives than beta-decays. They can take place when the total rest energy of the decay products (including the alpha-particle) is less than that of the original nucleus—in other words, when it is possible to increase the mean binding energy. The atomic mass number is reduced by 4 and the atomic number by 2 in an alpha-decay. A radioactive series consists of all the elements that are linked by alpha and beta radioactive decays. Among the heavy elements there are four such series.

Now turn to the Self-Assessment Questions and attempt numbers 5, 6 and 8 (pp. 50 to 51).

31.5 Power from Nuclei

31.5.1 Nuclear fission

This is the other important form of instability which turns heavy nuclei into more tightly-bound, lighter nuclei.

α-decay is really just a special case of *fission*, but it is such a special case that we have treated it separately. α-decay is the most important form of fission which happens spontaneously to the nuclides found in nature. Other forms of fission, the kind we actually refer to as 'nuclear fission', happen to certain heavy nuclei, especially when they have been excited in some way. It has been mentioned already that excited nuclei often emit γ radiation, but large nuclei frequently undergo fission instead.

fission

Nuclei may be excited in a number of ways, but one of the most effective is when a nucleus captures a slow neutron. The most practically important process of fission, up to the present day, is caused by the capture of a slow neutron on uranium $^{235}_{92}$U. You will remember that neutrons can get right up close to nuclei without having very much kinetic energy.

Why?

They are not repelled by the long-range electrical force which would repel a proton.

What is the result when $^{235}_{92}$U captures a neutron?

Excited $^{236}_{92}$U.

The resulting excited nucleus is a very unstable structure which oscillates rapidly between different shapes.

At this stage, we can introduce a model of nuclear behaviour which is, at once, utterly simple and quite accurate for heavy nuclei—*the liquid-drop model*.

liquid-drop model

BEFORE READING ON, DO HOME EXPERIMENT 1.

You have seen how a large liquid drop needs very little disturbance to break it into pieces. A heavy nucleus has some properties similar to those of a liquid drop.

What do the forces inside liquids and large nuclei have in common?

Both are held together by a short-range force. The nucleons, or molecules, at one side cannot feel the binding force of the nucleons, or molecules, on the other side—they can only feel the binding force due to their near neighbours. In liquids this gives rise to an important effect known as *surface energy*. An elongated drop can find a more stable, lower energy state by breaking into two smaller drops. This gives a reduction in the

surface area compared with the elongated condition. A reduction in the surface area reduces the surface energy. Surface energy is generally thought of as a macroscopic, i.e. large scale, effect as opposed to effects that are peculiar to the sub-microscopic world of the atoms and molecules. Similar 'macroscopic' effects apply to large nuclei where over 200 nucleons are jostling together in a small volume. 'Macroscopic' is in quotation marks here because nuclear sizes are still extremely tiny. The important point is that the diameter of a heavy nucleus is ten times as large as the range of the nuclear force. The capture of a neutron by $^{235}_{92}U$ is sufficient to set up a vibration in the resulting $^{236}_{92}U$ which distorts the nucleus so much that it can pinch across the middle—just as a liquid drop does when it is extended. The nucleus even assists in its own destruction in a way that a liquid drop cannot.

Can you guess how?

The two parts of the nucleus, as they begin to break away, will feel a smaller and smaller short-range attraction for one another, (due to the strong interaction) just as the two halves of a liquid drop would feel a smaller intermolecular force. But the long-range electrical repulsion, due to the protons in each part, will continue both during and after fission. Liquid drops do not have this additional factor tending to split them up. A further discussion of the liquid-drop model and quantum theory can be found in Appendix 2 (Black).

In the Home Experiment you may have noticed a few tiny drops left as the elongated large drop pinched into a column and then into two medium-sized drops.

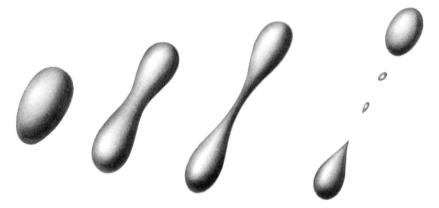

Figure 11 *The disintegration of a liquid drop.*

There is a nuclear analogy for these small drops too.

Can you guess what?

When nuclear fission occurs a few spare neutrons are produced in the fission process. Others are also emitted by the 'daughters' afterwards.

Why is it necessary for neutrons to be released?

You will remember from Figure 6 (c) that the heaviest nuclei are the ones with the largest proportional neutron excess. When two medium-sized

nuclei are produced by fission one of them might be a stable isotope—say we have for example,

$$^{236}_{92}U \rightarrow {}^{56}_{26}Fe + \text{another nucleus.}$$

What is the other nucleus?

If only two nuclei were produced, the other must be $^{180}_{66}Dy$. But the stable isotopes of $_{66}Dy$ have an atomic mass number A equal to about 162. There are 18 or so neutrons too many. Most of them are emitted from the daughter nuclides immediately as decay products, but some of them are delayed a short time. The other excess neutrons will change to protons with β^--emission. Thus, after fission has taken place due to the capture of one neutron, many more neutrons will be released. Because of detailed differences in the properties of the different uranium isotopes, these neutrons are usually too energetic to give immediate capture on $^{235}_{92}U$, but they may capture on $^{238}_{92}U$ and induce a fission of $^{239}_{92}U$. With this build-up of neutrons, it is clear that one fission can lead to more than one subsequent fission. This self-perpetuating sequence of events is called a *chain reaction*. An electrical chain reaction occurs when the input microphone of an amplifier is placed in front of one of the output speakers. An inaudible noise will be picked up and amplified so that it is picked up again, louder and so on. This 'positive feedback' rapidly builds the noise up to a deafening howl. A nuclear chain reaction may begin if:

(a) there are enough heavy nuclei (a *critical mass*) packed close together, to catch one another's neutrons. (There must be some spontaneous neutron-producing processes to get things started. The α-particles from spontaneous decays will often interact with other heavy nuclei nearby, and sometimes knock out neutrons. Such neutrons and those from β^-- and γ-scattering are sufficient to start a chain reaction.)

(b) the neutrons have the correct energy to be captured readily on the heavy nuclei provided.

chain reaction

critical mass

What are the two major technological applications of chain reactions?

They are the nuclear bomb (the 'A-bomb') and nuclear electric power stations. In the former, a *critical mass* of $^{235}_{92}U$ (or a similar nuclide) is brought together and an uncontrolled chain reaction is set up. The temperature of the interacting uranium rises very rapidly due to ionization and atomic excitation, both by the energetic daughter nuclei and in turn by their decay products. This causes an explosion which is well known to be far more energetic and destructive than any chemical explosive. Vast quantities of fast neutrons are generated during the short time before the bomb disintegrates, and it is the fast neutrons which cause the chain reaction in this case. A large fraction of the uranium is involved in the fission process and used up. Something over 125 MeV is released per fission.

Calculate the energy release in joules when 235 grams of uranium $^{235}_{92}U$ is used up.

There are 6×10^{23} atoms of $^{235}_{92}U$ in 235 grams—one gram atom. If each of these gives up 125 MeV, there are $6 \times 125 \times 10^{23}$ MeV released, i.e. $6 \times 125 \times 10^{23} \times 1.6 \times 10^{-13}$ joule $= 1.2 \times 10^{13}$ joule

One way to appreciate the amount of energy involved in 1.2×10^{13} joule is to express it in terms of electrical energy. One watt is the power used when one joule of energy is consumed every second. A fair sized factory might use about 4 megawatt (4MW) so 1.2×10^{13} joule would keep it going for a month.

In nuclear reactors such as those used in the present generation of power stations, the property of $^{235}_{92}U$ to capture slow neutrons, with subsequent fission of $^{236}_{92}U$ is used.

Can you guess how the fast neutrons from one fission process can be slowed down enough to be captured by $^{235}_{92}U$ again?

Figure 12 *Laying the graphite core of a nuclear reactor.*

Figure 12 is a photograph of the graphite blocks forming the core which houses the uranium-bearing fuel rods of a reactor in a power station. The most abundant isotope of carbon (graphite) is $^{12}_{6}C$, which has a particularly stable nucleus. This nucleus has no affinity for extra neutrons, so a high-energy neutron entering the graphite will tend to make a series of collisions with the carbon nuclei, giving up energy to them. The neutron is not absorbed, and in most cases it finally diffuses back with a suitably low energy into a fuel element.

The graphite, which slows down fast neutrons without absorbing them, is called a *moderator*.

moderator

How do you think the chain reaction can be controlled?

To control the chain reaction the number of free neutrons in the reactor core at any time must be kept constant. This is done by inserting *control rods* into the core. These contain nuclei with an affinity for neutrons. One such nucleus is boron $^{10}_{5}B$. When it captures a neutron it splits up, to form

control rods

32

the stable nuclei $^4_2$He and $^7_3$Li, so the neutron is lost forever. The neutron flux is monitored continually. Whenever it rises above a predetermined level, the control rods are pushed a little further into the reactor. The extra neutron absorption reduces the number of neutrons and the chain reaction proceeds at a lower level. Fortunately, as mentioned earlier, some of the neutrons are only emitted from the daughter nuclides after a delay of a fraction of a second. This is vitally important because it gives time to move the control rods.

31.5.2   The costs of nuclear power

Since 1942, when Enrico Fermi first observed a chain reaction, the technology of nuclear reactors has been steadily refined and developed. In Britain and elsewhere tremendous resources have been made available for this research. As a result the Central Electricity Generating Board is now faced with a very even choice between oil-burning and nuclear power stations. Nuclear and oil-fired stations at present being built are expected to produce electricity at a little over 0.2p* per kilowatt hour.** The cost of distributing the electricity will however, always be greater than this.

The coal-fired stations which are due to start work in the next two or three years are expected to have a generating cost of 0.3p per kWh. The cost of nuclear power has come down significantly from around 0.4p per kWh in the past ten years, but is likely to be kept up near its present levels by a number of factors, some of them economic and technical, others of a social nature. The chief social factor is the need to ensure that the radioactive materials from a reactor are not released in dangerous quantities. Radioactive materials in the air, in water supplies or in the sea can cause grave damage to human life and to the balance of nature. The safety of the community and the environment must be considered at every stage of nuclear power station design. The need for rigorous safety precautions adds to the capital cost of nuclear power plants, which is somewhat higher than for conventional stations. A 1300 MW oil-fired station, which the CEGB is now planning (1970), will cost £80 million to build. A comparable 'Advanced-Gas-cooled Reactor' nuclear station will cost £100 million. (However, of this £100 million, something like £16 million will be the cost of the initial charge of fuel. This fuel will keep the station going for a considerable period. One of Britain's nuclear stations removed the last of its original fuel rods in 1970 after ten years of operation. Each ton of fuel can give about 3 000 MW-days of power.) The cost of fuel rods is also influenced by the need for safety. Most of the radioactive isotopes produced within the nuclear reactor are contained within the material of the rods. When a rod has been used it must be treated chemically to extract and separate the useful nuclides*** that have been generated from those that are not only useless but dangerous. The most dangerous, with radioactive half-lives of tens or hundreds of years, are embedded in glass or concrete and are buried in deep mines or dropped to the deepest part of the ocean bed. The failure of one of these containers could lead to a dangerous release of activity and it is difficult to be sure that this will not occur for hundreds of years. There is also a certain amount of radioactivity released directly into the atmosphere by nuclear power plants. However this is generally very small and it is fairly easy to keep a continuous close watch on the amount of this contamination and shut down the reactor in the case of an unexpected increase. Despite these problems it is

---

* *Costs in this section are based on data from the Central Electricity Generating Board, November, 1970.*

** *1 kilowatt-hour (kWh) = 3.6 × 10⁶ joule.*

*** *The benefits to medicine and industry from artificial radioactive substances are many and varied, but are beyond the scope of this Unit to describe.*

expected that nuclear power plants will continue to show increasingly favourable generating costs, compared with conventional stations, largely because the cost of conventional ('fossil') fuels is bound to increase in the next few decades as they get scarcer. This trend is now seen in the U.S.A. and elsewhere, as well as in Britain.

31.5.3 Another source of nuclear power—fusion*

Take another look at Figure 9 (p. 25). You will remember that fission processes involve moving from the extreme right of the curve, up the slope towards the broad region of tightly-bound nuclei around $^{56}_{26}$Fe.

Where else would it be possible to move up a slope on this curve? What kind of processes would achieve such a movement?

Clearly there is a slope to move up on the left of the graph too. This would involve building up complex nuclei from simple ones with low values of the atomic mass number A. But we know that nuclei have positive charge, so nuclei which are close together repel one another. It is therefore difficult to push a pair of low mass nuclei very close to one another.

Why do they need to be very close to one another if they are to coalesce?

The short range of the strong interaction means that the two nuclei must be only around 10^{-15} m apart before they will attract one another. Kinetic energy is needed to bring the two nuclei this close together. This energy can be acquired in one of two ways. The first way is to take individual nuclei and accelerate them one by one. This is an important technique, but the number of nuclei that can be accelerated at once is too small for it to be useful in this context.

How else can nuclei be given very high energies?

The other way to give energy to nuclei is to raise the material containing them to a very high temperature. In Unit 5 you saw that hot material contains atoms and molecules with large kinetic energies. If the temperature is raised to above 10^6 K, then there will be some nuclear collisions with enough energy to overcome the effects of electrical repulsion. The nuclei will sometimes approach closely enough to feel one another's strong attraction, and *fusion* may take place.

fusion

The only artificial fusion process so far achieved is in the 'H-bomb'. In the early versions, a uranium fission bomb was used to achieve a very high temperature. A jacket of light elements around the A-bomb is heated to sufficiently high energies for fusion to occur. Several processes could be used of which the simplest are:

$$^2_1H + {}^2_1H \rightarrow {}^3_2He + n + 3.2 \text{ MeV}$$

or, with equal probability,

$$^2_1H + {}^2_1H \rightarrow {}^3_1H + {}^1_1H + 4.2 \text{ MeV}$$

i.e. deuterium (heavy hydrogen) is built up to an isotope of helium plus a neutron, with the release of 3.2 MeV. Alternatively tritium, the heaviest

* *A useful book that you might like to refer to if you wish to follow up this subject is:* H. R. Hulme, Nuclear Fusion, *Wykeham*, 1970.

hydrogen isotope, is produced with a proton and 4.2 MeV kinetic energy.

A deuterium-tritium reaction occurs at lower temperatures, but tritium is very expensive.

However, fusion processes are ultimately the source of all available energy on the Earth, in that they produce stellar energy. The centre of a star, such a star as the Sun for instance, contains compressed material at a very high temperature. If the temperature were not so high, it could not support the enormous mass of material piled on top of it. Remember the pressure in a gas depends on the temperature. At the centre of the Sun the material forms a very hot gas called a *plasma*. In a plasma there are very few neutral atoms. The individual kinetic energies of the atoms are much higher than the binding energies of many of their electrons. Whenever there is a collision between two such atoms, they tend to lose their electrons. A plasma contains lots of such free electrons, and atoms in the form of positive ions. They are all moving with very high velocities and undergoing repeated collisions with one another—electrons with electrons, electrons with ions, and ions with ions. If atoms have been completely stripped of their electrons by earlier collisions, then an ion-ion collision is really a collision between two bare nuclei.

plasma

How will such a situation give rise to fusion processes?

In exactly the same way as in the H-bomb—which produces a plasma when it is detonated. It is known that the Sun is largely composed of hydrogen, with some other light elements, but very small quantities of heavy elements. In the core of the Sun, the products of a particular nuclear encounter are not carried away rapidly from the highest temperature region. This region is very large—thousands of kilometres across. If a certain nuclide is produced in one encounter, it may, at a later time, take part in another. There can therefore be a number of other fusion processes as well as the one-step processes, such as the two examples we gave for possible 'H-bomb' reactions.

One sequence of reactions which is thought to occur in the Sun is shown in Figure 13.

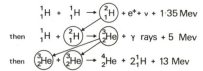

This, and other, 'hydrogen-burning' processes provide the source for the enormous amounts of solar energy which are radiated to us and into space. These fusion processes only occur in the centre of the Sun. A whole sequence of heat-transport mechanisms carries the energy up to the surface regions before it is radiated to us as visible light.

Figure 13   A sequence of possible fusion reactions inside the Sun.

Attempts have been made to build apparatus in which controlled fusion processes will occur. So far they have not been successful because our techniques for containing very hot plasmas are not good enough.

Why cannot plasmas be contained in ordinary steel retorts?

If a plasma touches a solid object, the solid is vaporized by the high temperature of the plasma, and the plasma itself cools down rapidly. So it is necessary to invent some other means of restraining the movement of the plasma; this is provided by a suitable arrangement of magnetic fields. We saw just now that a plasma is composed of charged particles, positive ions and electrons—all moving at high speeds. A fast moving charged

particle is in itself an electric current. But electric currents are deflected sideways by magnetic fields.

Such deflection of electric currents by magnetic fields is the basis of the operation of electric motors in anything from hair dryers to tube-trains and of the usual kind of loudspeaker. As you saw in the TV programme of Unit 6 this effect can be demonstrated by putting a magnet close to the screen of a television set. *Do not do this if you have colour TV.* As the electrons in the tube approach the screen, they are deflected by the magnetic field close to the magnet.

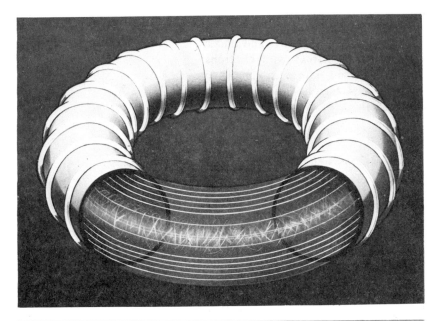

Figure 14

(a) *A magnetic means of containing a plasma.*

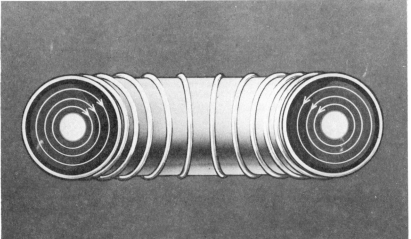

(b) *Showing the secondary magnetic field in the 'plasma bottle'.*

One lay-out of magnetic field that has been used as a magnetic bottle for plasma is shown in Figures 14 (a) and 14 (b). The doughnut-shaped field forms a 'bottle' that has no end. Charged particles in the plasma might start off in an outward direction from the ring, but they deflect sideways around the field direction and cannot get away. The curved lines on Figure 14 (a), running across the cut-away part of the tube, represent the direction of the magnetic field. The tangle of fine lines at the core of the tube represents the plasma of ions spiralling around the field lines. Figure 14 (b) shows the secondary magnetic field produced when a current is induced in the plasma. This secondary field pinches the plasma in towards the core of the tube. The induced current both heats the plasma and causes it to pinch away from the walls of the bottle. If high enough temperatures

could be attained for long enough, the fusion process itself would heat the plasma, producing a fusion chain reaction.

The trouble with making a long-lasting bottle of hot plasma is that the plasma wanders away from its proper place and hits the walls of the containing vessel. So far, all attempts to contain plasma have worked for too short a time for the temperatures to be raised to the level required for fusion to begin.

Research on this problem has been pursued vigorously throughout the last decade, but success is coming very slowly. If fusion processes could be made to occur in a controlled way, and if a way could be found to convert the energy produced into heat, the world would have a new energy source with nearly unlimited supplies of fuel.

Where is this fuel?

Controlled fusion processes will probably use deuterium as a fuel and possibly hydrogen or lithium. The seas contain sufficient of these to satisfy all our foreseeable energy requirements. This is in sharp contrast to world supplies of oil and coal. At the expected rate of consumption, based on current trends, world reserves of these fossil fuels could be exhausted within one hundred years. The long-term situation is more serious, because oil and coal form a unique reserve of organic chemical raw materials. A substantial amount of them must be held for these purposes, and not simply burnt.

31.5.4 Summary of section 31.5

Large nuclei such as uranium 235 can be made to capture an extra neutron and then break into two fragments of smaller size. This process is called nuclear fission and is in principle similar to alpha-decay. It can occur when the total rest energy of all the fragments is less than that of the original nucleus—uranium 236 in this case. The breaking of a liquid drop into two smaller drops is a useful model of the fission process (the liquid-drop model). Nuclear fission releases large amounts of energy and is the basis of nuclear power stations and the atom bomb. Neutrons released from a fission are captured and cause further fissions. If at least one neutron from each fission goes on to produce a further fission, a chain reaction may take place which is uncontrolled in an atom bomb but controlled in a nuclear power station. The control is by neutron-absorbing control rods which are withdrawn from the reactor core just far enough to allow a chain reaction to develop; the core also contains a moderator to slow down the neutrons so that they can be captured more easily by the uranium. Nuclear stations can produce electricity appreciably more cheaply than modern coal-fired stations. Rigorous precautions are necessary in nuclear power stations to prevent the escape of radioactive material into the surrounding area. This is discussed further in Unit 34.

Energy can also be released when two light nuclei such as hydrogen are fused together—nuclear fusion. This progress requires very high temperatures and is the basis of the hydrogen (H) bomb. Such temperatures are also found inside stars such as the Sun; fusion is the source of the energy for the Sun's radiation. Attempts to control a fusion reaction for use in power stations by confining a plasma in a magnetic field have so far been unsuccessful.

Now turn to pp. 50–1 and attempt the remaining Self-Assessment Questions, 4, 7, 9 and 12.

Summary of the Unit

The nucleus of an atom has a diameter of a few femtometres—some 10^5 times smaller than the atom itself. This diameter can be measured by scattering a beam of high momentum particles, such as electrons, from the nuclei and analysing the distribution of scattered particles. The nucleus is held together by a force called the 'strong interaction' having a very short range—about 2 femtometres. The values of the binding energy of the first nucleon in various nuclei indicate an energy-level structure for the nucleons, analogous to the energy levels of atomic electrons. Further evidence about nuclear structure is obtained from nuclear spectroscopy, in which the wavelengths of gamma-rays (photons of very short wavelength) from excited nuclei are measured. Two separate energy-level structures (shell structures) are indicated, one for the protons and one for the neutrons. For a given atomic mass, only a very few nuclides are stable; for lighter elements, the numbers of neutrons and protons are about equal, but heavier elements have a neutron excess. A nucleus with the wrong relative number of neutrons and protons is unstable; it becomes stable by radioactive decay, converting protons to neutrons by β^+-decay or neutrons to protons by β^--decay. Alternatively some nuclei, especially among the heavy elements, undergo α-decay. This is a special case of nuclear fission. In other fission processes, a heavy nucleus breaks up into two smaller pieces, usually of unequal but comparable sizes. When fission occurs, considerable amounts of energy are released in the form of kinetic energy of the fragments. A self-perpetuating chain-reaction can occur, which is either uncontrolled (A-bomb) or controlled (nuclear electric power station). Nuclear power stations are likely to produce a large proportion of the world's electricity in the forseeable future. Energy is also released by the fusion of light elements such as hydrogen. Fusion occurs in stars and the 'H-bomb'; research aimed at controlling a fusion reaction to generate power steadily and continuously has so far only been partially successful, but success in this would probably satisfy all forseeable energy needs of the world.

Appendix 1 (Black)

Tunnelling—how quantum theory affects α-decay

What do $^8_4$Be, $^{12}_6$C, $^{16}_8$O, and $^{20}_{10}$Ne have in common?

Each of these nuclei has in it two, three, four or five times the number of neutrons and protons that there are in $^4_2$He. The nuclei $^{12}_6$C, $^{16}_8$O, and $^{20}_{10}$Ne often behave like three, four or five α-particles bound together. We have already mentioned that $^8_4$Be very rapidly becomes two α-particles.

Why do you think these nuclei behave like composites of α-particles?

The reason is simple. The α-particle is a very stable, and relatively small, structure. Inside a large nucleus, nucleons tend to pack together in the snuggest possible way. One way in which they might achieve this is if they combine in the form of α-particles inside the larger nuclear structure. There is experimental evidence that such α-particles do exist in nuclei, at least near the surface.

You may have some difficulty at this point switching from one model of the nucleus to what is essentially another. Up to now we have taken the atomic energy-level structure as our model for what is going on inside the nucleus —and a very successful model it is too. But like all models and analogies they must not be pressed too far. There *are* differences between the forces acting on electrons in an atom and the forces acting on the nucleons inside a nucleus. In an atom there is a well-defined centre of attraction (the nucleus); in the nucleus itself, because of the short range of the strong interaction, there is no well-defined centre of attraction for the nucleons. We could therefore expect nucleons to show some tendency to form compact localized clusters (α-particles). When one comes across a phenomenon that is extremely closely related to this particular behaviour of nucleons, it is fruitless to try and describe it using a model not adapted to that kind of situation. It is best to change to another mental picture—one in which the nucleus is regarded for at least some of the time as being made up of α-particles in motion rather than individual nucleons in motion.

Some heavy nuclei emit α-particles in order to achieve a more stable arrangement of their constituents. If one wishes to understand α-decay on a deeper level, it helps to use this picture of α-particles already existing inside the heavy nucleus before the decay happens.

Look back to section 31.4.2. What is the binding energy of the first α-particle in $^{238}_{92}$U?

Apparently this must have a negative binding energy! You would have to *give* an α-particle 4.2 MeV to get it *into* $^{234}_{90}$Th nucleus, to make $^{238}_{92}$U. Binding energy is defined as the energy needed to get a particle *out of* a nucleus.

Clearly this is why $^{238}_{92}$U decays into $\alpha + ^{234}_{90}$Th, but it takes a very long time to do it. Its half-life is 4.51×10^9 years. So one can picture the α-particle trying for something like four and a half thousand million years before it

is permitted to do something which it might be expected to do immediately; that is, to jump out and get away with 4.2 MeV of kinetic energy.

To understand what causes the delay one needs to use a little quantum theory, and to introduce the concept of a *potential distribution*.

NOW DO HOME EXPERIMENT 2 (BLACK). THIS WILL SHOW YOU THAT A POTENTIAL DISTRIBUTION IS QUITE AN EASY THING TO PICTURE.

What does the decay α-particle in $^{238}_{92}$U have in common with a marble on an upturned plate?

Quite a lot, since both have an excess of potential energy. When either particle 'escapes' from its system, this excess of potential energy appears as kinetic energy. Once the particle leaves the nucleus it does so with 4.2 MeV of kinetic energy. Similarly, if one nudges the marble on to the slope of the rim it will run off across the carpet.

But the potential distribution for the α-particle is a more complicated shape than shown in Figure 3 in Home Experiment 2. Within a few fm of the centre of the nucleus the α-particle will feel a strong attractive force.

Let us try to build up a picture of the potential experienced by an α-particle in the region of a heavy nucleus. For a start, the charge on the nucleus is not all concentrated at one point, so the electrostatic part of the potential does not look like Figure 3.

Sketch the electrostatic potential distribution you expect near and within a large nucleus.

Your sketch should look like Figure 15.

Because the charge on the protons is distributed throughout the volume of the nucleus, there is no sharp increase in the potential when $r \approx 0$ at the very centre.

In 31.3.1, you learned that a nucleon deep inside a heavy nucleus is acted upon by all the surrounding nucleons in a balanced way, but at the edge of the nucleus (Figure 2, p. 14) a nucleon will be attracted inwards by the strong interaction of its neighbours. An α-particle is made up of nucleons, so it would also be attracted inwards when it is near the nuclear edge.

Sketch the strong interaction potential distribution you expect for a nucleon or α-particle near or within a large nucleus.

Your sketch should look like Figure 16.

Notice that this is approximately a 'square well' potential. The similarity with a hole in a putting green is not fortuitous. A particle remote from the nucleus feels no force—there is no slope to the potential. A particle deep inside the nucleus feels no force, just as the golf-ball may lie anywhere at the bottom of the hole. But the steep sides have an inward slope which corresponds to a strong inward force on a particle at about 5 fm from the centre, like a ball running down into the steep-sided hole.

An α-particle experiences both electrostatic and strong forces.

Sketch the shape that you would expect the effective combined potential to have for an α-particle in or near a large nucleus.

Figure 17 gives the shape you should have sketched.

You will notice that the strong interaction potential dominates the potential within the nucleus. The electrical potential dominates outside.

At point A it is in the middle of the potential well, so its potential energy, U, is -10 MeV. Its kinetic energy, T, must therefore be $+8$ MeV so that $U + T = -2 = E$, the energy of the level (notice that $-E$ is the same as the binding energy of the α-particle).

At point B, the potential energy, U, is -2 MeV, so the kinetic energy T, is zero. One can compare the motion of the α-particle inside the potential well to a marble rolling around *inside* a plate *right way up*. The highest point the marble can reach on the rim of the plate is the point at which all of its kinetic (rolling) energy has been turned into potential energy (height). When it runs back onto the bottom of the plate, it gets the kinetic energy back again.

If Figure 17 represents the potential distribution for α-particles in $^{238}_{92}$U, draw in the energy level for the α-particle which waits 4.51×10^9 years or so before escaping.

When you have done this (and *not* before), check your energy level with Figure 18 on p. 43.

On Figure 18 you see what you should have drawn. Notice that the level energy in this case is 4.2 MeV above zero.

What seems to be stopping the α-particle from getting out?

The α-particle cannot get out because there is a potential hill between the nucleus and the outside world. The downward slope runs inwards on the inside of this hill.

Why is the downward slope directed inwards on the side of the potential hill that is closer to the centre of the nucleus?

An inward slope denotes an inward force. The strong attraction of the nuclear force pulls the α-particle back when it tries to leave the edge of the nucleus. This strong attraction dominates the electrical repulsion at the inside edge of the barrier region.

In Figure 2 (p. 14) you saw how a nucleon at the edge can feel an attractive force, since all the other nucleons are attracting it from one side only. The same is true for an α-particle which tries to leave the edge of the nucleus. This is what the inward slope of the potential distribution represents.

Why is the slope directed outwards on the side of the potential hill that is further from the centre of the nucleus?

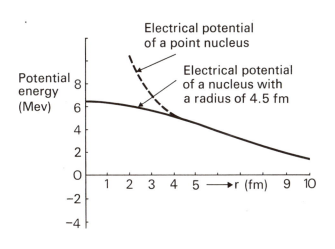

Figure 15 *Electrical potentials of a point nucleus and of a large nucleus.*

Figure 16 *Strong interaction potential for an α-particle due to a large nucleus.*

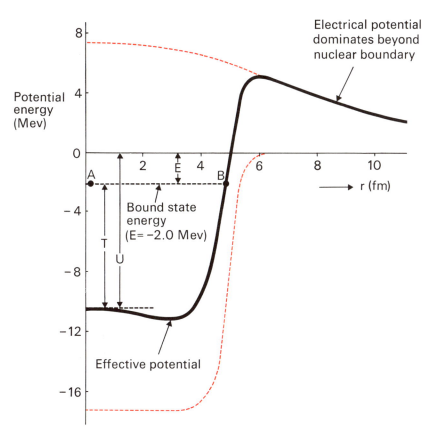

Figure 17 *Effective potential for an α-particle in and near a large nucleus.*

Once the α-particle has left the short range of the strong interaction, it feels only the electrical repulsion.

So now one can see what the problem is. In terms of 'common sense' physics, there is no way in which an α-particle can jump over the potential hill and get out. It is like a zoo. The animals are often kept in cages with open tops and high slippery sides. If a lion got out of its cage, then it

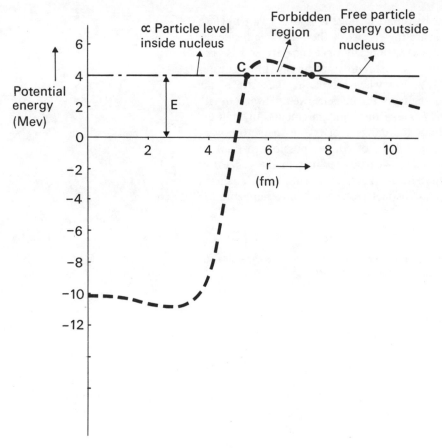

Figure 18   *The potential barrier for an α-particle.*

would be free to run off and go anywhere it chose. But a lion cannot spring over the side of its cage. It can never have sufficient kinetic energy to get over the potential hill of the cage side. α-decay of $^{238}_{92}$U appears just as impossible as a lion jumping a twenty-foot wall, but it happens. The question is not 'What delays the α-particle from being emitted?', which is the problem faced earlier in this section, but rather 'How is it possible for the α-particle to be emitted at all?'!

The reason has to do with Heisenberg's uncertainty principle.

How can the uncertainty principle help?

As you saw in Unit 29 there is an uncertainty relation that can be expressed mathematically as $\Delta E . \Delta t \approx h$. That is, during a short time Δt, we cannot be sure of the energy of a particle to better than some precision ΔE, where $\Delta t \approx h/\Delta E$. Now the whole trouble with getting over the potential hill is that, in the region from point C to point D of Figure 13, the value of the potential energy U is greater than the level energy E. If a particle is to get out of the nucleus therefore, it must somehow increase its energy for a short time—the time required to get from C to D.

Is there anything to stop it having a fluctuation in energy?

The law of conservation of energy says that if a long period is available for measuring the energy of a system, then successive readings each taken

over long periods of time will be found to have a constant value. Thus, we can take our time and measure the energy of the α-particle inside the nucleus very precisely prior to decay; similarly once the particle has been emitted we can take a long time to measure its final energy. In both cases the answer is precisely 4.2 MeV. But when discussing the state of the α-particle as it is actually emerging from the nucleus, we are trying to talk about something that happens *between* these measurements. In Unit 29, you were strongly warned about the dangers of trying to picture what happens in between measurements. It is, of course, possible to think of trying to measure the energy of the α-particle as it is in the process of passing through the potential barrier. Then we would once again be talking about a measurement as distinct from something happening in between measurements.

But would such an intermediate measurement tell us that the α-particle still had energy 4.2 MeV as it passed through the barrier?

Probably not. This measurement of energy would have a precision no better than that given by the uncertainty relation. There is therefore no experimental evidence that the energy of the α-particle (or indeed of anything else) must remain constant from one short interval of time to the next. But earlier, when we spoke of the 'impossibility' of the α-particle getting out of the nucleus, the argument was based on the belief that the energy of the particle had at all times to have a precisely specified energy of 4.2 MeV. According to the uncertainty relation, this belief is quite unjustified. There is nothing in fact to stop us using theoretical models in which the α-particle is considered capable of 'borrowing' an energy ΔE for a time Δt given by the uncertainty relation.

Unfortunately for the α-particle, the rather large amount of energy ΔE needed means that its chance of getting out in the time Δt is very small indeed. But it keeps on trying, and after 4.51×10^9 years there is a 50–50 chance that it will have succeeded.

Do you expect that the chances of the α-particle getting out of a particular nucleus will change with time?

You saw in Unit 6, section 6.3, that there is no change of the decay probability with time. In a sample of $^{238}_{92}$U, or any other of the nuclides listed in Unit 6, the rate of α-emission of the sample depends on

What?

It depends upon the kind of nuclide, in particular its half-life, and the amount of it present in the sample.

The process by which an α-particle escapes from a potential well, through a region of high potential, is called *tunnelling*. Such tunnelling through *potential barriers* occurs in other contexts in nuclear physics, and in semiconductor devices used in electronics.

tunnelling
potential barriers

The liquid-drop model and quantum tunnelling

In a detailed calculation of the speed of a fission process, using the liquid-drop model, physicists must also consider quantum tunelling effects. The excited $^{236}_{92}$U nucleus sits there shaking for a while like a wobbling liquid drop, after the slow neutron has been absorbed. But the strong nuclear force causes a potential barrier against break-up in this case, just as we saw it did in α-decay. To pinch across the middle and break, the nucleus needs to borrow a little energy for a short time. In this case, the energy involved is small enough that the process can go ahead very rapidly— fast enough to beat the alternative way that the excited nucleus can lose energy, by γ-ray emission. It doesn't have to wait millions of years.

> **Do you expect quantum tunnelling effects to be significant in the break-up of real liquid drops?**

The value of h is so small that in the macroscopic world we rarely notice quantum effects. $h = \Delta E.\Delta t = 4.0 \times 10^{-21}$ MeVs. This is quite a big number on the nuclear level where 4.0 MeV is a typical binding energy and things often happen in 10^{-21} second. In macroscopic units, $h \approx 10^{-34}$ Js. The surface energy of a small drop of water might change by 10^{-5} joule when it is split up. But in 10^{-29} second the two parts could not get very far apart!

Appendix 3

Airships and alpha-particles

The Zeppelins and other airships from the First World War, until the 1930s, were filled with hydrogen gas. With only thin canvas between the hydrogen and the air, the danger of fire was enormous. In fact, disastrous fires wrecked many of the largest airships. None have been used commercially since the mid-30s for that reason.

What other gas could be used to fill a lighter-than-air balloon?

Already in the 1930s it was known that helium gas would be a non-inflammable substitute for hydrogen in airships. Unfortunately, helium is expensive and there are very few sources of pure helium. This may seem strange, since the helium nucleus is so stable, and there are so many α-emitting nuclei in the world. In fact, its great stability means that ^4_2He is the second most abundant nuclide in the whole universe. Why then is there so little of it available on Earth?

The answer lies in two other properties of helium—its low chemical affinity for other atoms and its lightness. Since helium has closed electron shells, it does not bind very readily into compounds. It prefers to float free as a gas, and since it is lighter than air, it floats right to the top of the Earth's atmosphere. The gas at the top of the atmosphere is exposed to the full radiation of the Sun. This heats it to high temperatures and molecules of a high temperature gas have high kinetic energies (see Unit 5, section 5.2.3). For helium the kinetic energies involved correspond to such high velocities that individual molecules can completely escape from the Earth's gravitational attraction and drift off into space. This was discussed in Unit 27.

Why do the Earth's hydrogen, oxygen and nitrogen not diffuse away in the same way?

In fact hydrogen does, but so much of it is bound up chemically in water that we will never run out of it. Oxygen $^{16}_8\text{O}$ and nitrogen $^{14}_7\text{N}$ both are heavier than helium ^4_2He, and they form diatomic molecules, while helium is monatomic. This means that O_2 molecules are approximately $2 \times 16/4 = 8$ times as heavy as He.

N_2 molecules are how many times heavier than He?

Approximately,
$$\frac{2 \times 14}{4} = 7 \text{ times heavier than He.}$$

The temperature of the upper atmosphere determines the range of values of the kinetic energy T of the gas molecules. The velocity of one such molecule is then found from equation 25 of Unit 4, section 4.4.4:

$$T = \tfrac{1}{2} mv^2$$

$$\therefore \quad v^2 = \frac{2T}{m}$$

$$\text{so } v \propto m^{-\frac{1}{2}}$$

At a given temperature, the range of kinetic energies is the same for all molecules whatever the value of m.

What is the ratio of velocities between an oxygen molecule and a helium atom if both have the same kinetic energy?

The mass of an atom of helium is approximately 4 units and of a molecule of oxygen 32 units, so the ratio of masses in about 8:1. Hence the ratio of velocities is $1:\sqrt{8}$ or $1:2.8$. An oxygen molecule moves more slowly than a helium atom by this factor. For nitrogen the ratio would be 1:2.7.

This difference in velocities is enough to allow the Earth's gravity to hang on to our oxygen and nitrogen, though our free helium can escape. So where does North Sea gas come in?

North Sea gas comes out of the ground. It has been there for a long time, gradually collecting under an impervious layer of rock. Among the minerals under that rock are some containing members of the radioactive series. Each α-particle from a decay captures two electrons and becomes an atom of helium. Some helium diffuses out of the material in which it was formed and collects with the petroleum gas under the impervious layer of rock. When the gas is released the helium comes with it. This helium can be extracted by commercially viable processes, so it is now possible to think realistically of filling commercial airships with helium. However, even the natural gas sources may begin to run out within a few generations.

Artificial nuclear reactions give off α-particles and it is amusing to see whether a nuclear-powered airship could make its own helium.

If such an airship is powered by a process that gives an α-particle as a by-product for every 6.125 MeV of energy produced, calculate the rate at which it can make up its own helium supply. Assume its motor generates one megawatt of power.
(Assume 1 MeV $= 1.6 \times 10^{-13}$ joule)

One megawatt is 10^6 joules per second

$$= \frac{10^6}{1.6 \times 10^{-13}} \text{ MeV per second}$$

The number of particles per second will be:

$$\frac{10^{19}}{6.125 \times 1.6} = 10^{18}$$

One mole of helium contains roughly 6×10^{23} α-particles (Avogadro's number). One mole of a gas at standard temperature and pressure occupies 22.415 litres, say 20 litres. Hence, in one second *approximately* $\frac{20 \times 10^{18}}{6 \times 10^{23}}$ litre of helium are produced.

In one hour about $\frac{60 \times 60 \times 20}{6 \times 10^5} = 0.012$ litre of helium are produced.

This is not enough to make up for the leakage, by diffusion, of helium from the enormous balloon.

Appendix 4

Glossary

CHAIN REACTION   A fission reaction initiated by a neutron, and in which sufficient neutrons are given out for the reaction to be self propagating.

CHARGE INDEPENDENCE   An interaction whose strength does not depend on the charge of the particles involved. This is an important property of the strong interaction.

CONTROL ROD   Rod made of a material which strongly absorbs neutrons. By varying its position in a reactor, the power level can be controlled.

CRITICAL MASS   The minimum mass of a given fissile material in which a chain reaction can occur.

EFFECTIVE NUCLEAR RADIUS   The radius of the nucleus as deduced from a particular experiment. Different types of experiments measure slightly different radii.

LIQUID-DROP MODEL   The analogy of fission with the break-up of a drop of liquid when disturbed.

MEAN BINDING ENERGY PER NUCLEON   The total energy needed to take a nucleus apart, divided by the number of nucleons.

MODERATOR   Material placed between fuel elements in a reactor to slow neutrons down without appreciable absorption.

NEUTRON EXCESS   The difference between the number of neutrons in a nucleus and the number of protons.

NUCLEAR FISSION   The splitting of a nucleus (usually a heavy nucleus) into two fragments of approximately half the size of the original nucleus, with the release of energy.

NUCLEAR FUSION   The formation of a stable nucleus by the addition of two lighter nuclei.

NUCLEON   A proton or a neutron (or their antiparticles).

NUCLEON DENSITY   The number of nucleons per fm³.

NUCLEON ENERGY LEVEL   One of the discrete series of allowed energies for a nucleon in a nucleus.

POSITRON   The positive counterpart (antiparticle) of the negative electron.

RADIOACTIVE SERIES   A series of unstable nuclei linked by α-decay.

STRONG INTERACTION   The dominant attractive interaction between nucleons at distances less than 1 fm.

TOTAL NUCLEAR BINDING ENERGY   The total energy needed to take a nucleus apart.

Question 1 (*Objective 2*)

Choose the correct alternative in each of the following statements.

(a) The radii of most nuclei are a few times $10^{-17}/10^{-15}/10^{-12}$ m.

(b) The radii of nuclei are much less/the same/much more than the range of the strong interaction.

(c) Every known nucleus has a radius of more than $10^{-14}/10^{-16}/10^{-18}$ m.

(d) A nucleon in a large nucleus feels the strong attraction of one/a few/all of the other nucleons.

Question 2 (*Objective 3*)

The first diffraction minimum for the scattering of neutrons from zinc $^{64}_{30}$Zn falls at 30°. The effective-radius of the zinc nucleus is given by

$$r = 1.2 \, A^{\frac{1}{3}} \text{fm}.$$

Calculate the approximate momentum of the neutron beam ($h = 6.6 \times 10^{-34}$ J s).

Question 3 (*Objective 1*)

Decide which of the following statements is a suitable definition of the following terms.

    A Mean binding energy per nucleon.

    B Neutron excess.

    C Nuclear energy level.

    D Excited nucleus.

1   A situation in which β^--decay is bound to happen.

2   An energy at which neutrons, or protons, may exist inside a nucleus.

3   The energy of the neutrons, or protons, in a nucleon beam.

4   A nucleus which is not in its ground state.

5   The nucleus of an ionized atom.

6   The difference between the number of neutrons N, in a nucleus, and the atomic mass number A.

7   The energy required to remove all of the nucleons from the nucleus, one by one.

8   The mass defect of the nucleus, expressed in MeV/c^2.

9   A situation in which there are more neutrons than protons in a nucleus.

10   The difference between the number of neutrons N, in a nucleus, and the atomic mass number A.

11   The energy required to break up the nucleus into its constituent nucleons, divided by atomic mass number A.

12   A nucleus that has just fallen to its ground state by γ-ray emission.

49

Question 4 (*Objective 1*)

Decide which of the following statements is a suitable definition of the following terms.

A Moderator.

B Nuclear Fusion.

C Control Rod.

D Critical Mass.

E Nuclear Fission.

1 The material which reduces the number of free neutrons in a reactor.

2 The mass of uranium, or other fissionable material, which is contained in one fuel rod of a nuclear power reactor.

3 A rod containing a nuclide with a special affinity for neutrons, which reduces the number of neutrons in a reactor when it is pushed deeper into the core.

4 A process in which two heavy nuclei with small total binding energies come together to form an even heavier nucleus with a larger total binding energy.

5 A material which is introduced into a reactor to slow down neutrons.

6 A carbon rod which is placed in a nuclear reactor to increase the number of free neutrons and encourage fission processes.

7 The amount of uranium, or other fissionable material, which, when it is contained in a suitably small volume, will undergo a spontaneous chain reaction.

8 The rod of uranium which is pushed further into the reactor when the temperature rises above a safe level.

9 A process involving light nuclei which increases the mean binding energy per nucleon.

10 That mass of water which, when formed into a freely forming drop, will break spontaneously into two or more smaller drops.

11 The block of a special type of boron which forms the fuel of a nuclear reactor.

12 A process in which a heavy nucleus, which may have been excited, splits spontaneously into two or more daughter nucleii.

Question 5 (*Objective 7*)

When a $^{20}_{11}$Na nucleus undergoes β-decay, do you expect it to emit an electron or a positron?

Question 6 (*Objective 6*)

Use the Figures 6 (a) and 6 (b) (pp. 18, 20) in the text to answer the following questions.

(a) How many stable nuclides have $Z = 14$?

(b) How many stable nuclides have $A = 50$?

(c) How many observed nuclides are shown with $N = 132$?

(d) Is $^{205}_{80}$Hg stable?

(e) Is $^{35}_{16}$S stable?

(f) Is $^{7}_{3}$Li stable?

(g) Does $^{28}_{95}$Am exist?

Question 7 (*Objective 7*)

Do you expect nuclides with odd or even numbers of neutrons to make good moderators for chain reactions? Do you expect nuclides with odd or even numbers of neutrons to make good control rod material?

Question 8 (*Objective 7*)

Using Figure 9 (p. 25) in the text, work out the Z and A values for the nuclide produced in the decay series from U^{238}_{92} after six α-decays and two β^--decays. What element is this?

Question 9 (*Objective 4*)

Why are high-energy particles needed to examine the structure and size of a nucleus?

A Because the electrons around the nucleus act as a shield.

B Because the nuclear charge repels charged particles aimed at it.

C Because the incident particles should have a small wavelength.

Question 10 (*Objective 5*)

Below, there is a list of binding energies for the first nucleon to be removed from a sequence of nuclides beginning at $^6_3$Li. Notice that each successive nuclide is the one which remains after removing the first nucleon from its predecessor. (Check that you understand the significance of a negative binding energy.)

(a) List in succession the nucleons which have to be removed from each nuclide to reach the next on the list.

(b) Calculate the mean binding per nucleon.

$^6_3$Li

       5.17 MeV

$^5_3$Li

     − 1.47 MeV

$^4_2$He

      20.5 MeV

$^3_2$He

      5.44 MeV

$^2_1$H

      2.18 MeV

$^1_1$H

Question 11 (*Objective 9*)

Using the list of binding energies for the first nucleon in question 10, calculate the mass defect of $^4_2$He in a.m.u. (Slide-rule accuracy is sufficient; use Table I (p. 19) in the text for any additional information.)

Question 12 (*Objective 10*)

There are three statements below about the merits of methods for generating electric power. Which of these statements are basically unfavourable to nuclear power generation by fission, by a fusion technique, or by both, or neither?

A   Major technical problems still remain unsolved.

B   The power stations will cause air pollution.

C   The fuel must be radioactive and therefore dangerous to prepare.

D   The method consumes a natural resource which is in limited supply.

E   The method creates dangerous waste materials.

Question 1

(a) 10^{-15} m (see the beginning of section 31.2.1).

(b) Much more (see Figure 2, p. 14).

(c) 10^{-16} m (this can be deduced from equation 3 (p. 10) with $A=1$).

(d) A few (see Figure 2).

Question 2

$^{64}_{30}$Zn is the convention for the nucleus with $A=64$; $Z=30$. So using the equation given:

$$r=1.2\ A^{\frac{1}{3}}$$
$$=1.2\ (64)^{\frac{1}{3}}=1.2\times 4$$
$$=4.8\ \text{fm.}$$

From equation 2 in the text,

$$\lambda=2r\sin\theta/1.22.$$

In the experiment $\theta=30°$, so $\sin\theta=0.5$.

Thus
$$\lambda=\frac{2r\ 0.5}{1.22}=\frac{r}{1.22}.$$

However
$$p=\frac{h}{\lambda}=h\times\frac{1.22}{r}=\frac{h\ 1.22}{4.8\times 10^{-15}}$$

so
$$p=\frac{6.6\times 10^{-34}\ 1.22.}{4.8\times 10^{-15}}$$

$$=1.65\times 10^{-19}\ \text{N s}$$

Question 3

A   Mean binding energy per nucleon: 11
B   Neutron excess: 9
C   Nuclear energy level: 2
D   Excited nucleus: 4.

Question 4

A: 5
B: 9
C: 3
D: 7
E: 12.

Question 5

The $^{20}_{11}$Na nucleus has $A = 20$, $Z = 11$, therefore the number of neutrons, N, $= 20 - 11 = 9$.
So, in this case, $Z > N$.
So you should *expect* a positron to be emitted.

Question 6

(a) $Z = 14$ corresponds to Si; 3 stable nuclides with $A = 28$, 29, 30.

(b) $A = 50$: 3 stable 'isobars' $^{50}_{24}$Cr; $^{50}_{23}$V; $^{50}_{22}$Ti.

(c) 8 observed nuclides have $N = 132$ from $^{214}_{82}$Pb to $^{221}_{89}$Ac.

(d) No.

(e) No.

(f) Yes.

(g) No.

Question 7

For a moderator there should be no strong affinity for an extra neutron. Even numbers of neutrons are desirable, as in $^{12}_{6}$C. There is one exception. Sometimes $^{2}_{1}$H is used as a moderator, but this is a special case. Being a light atom it is an excellent moderator. For a control rod there should be an affinity for an extra neutron, so an odd number is preferred, as in $^{10}_{5}$B.

Question 8

The end product is $^{214}_{82}$Pb; lead.

The new values of A and Z are worked out as follows:

$$
\begin{aligned}
Z \text{ (end)} &= Z \text{ (start)} - (6 \times 2) + 2 \\
&= 92 \quad - 12 \quad + 2 = 82 \\
A \text{ (end)} &= A \text{ (start)} - (6 \times 4) \\
&= 238 \quad - 24 \quad = 214
\end{aligned}
$$

Question 9

C The effect described in section 31.2.1 is a *diffraction* effect, which only takes place if the wavelength of the incident particles is comparable to the size of the nucleus.

Question 10

(a) n; p; n; p; n   (n = neutron; p = proton)

(b) Since mean b.e./nucleon = (total b.e.)/(number nucleons)
    Mean b.e./nucleon = (sum of energies given)/6
    $\qquad\qquad\qquad = 5.30 \ MeV/nucleon$

Question 11

The binding energy of the $^4_2$He nucleus is the sum of all the energies in question 10 up to $^4_2$He,

$$\text{i.e. b.e. of } ^4_2\text{He} = 2.18 + 5.44 + 20.5 \text{ MeV}$$
$$= 28.12 \text{ MeV}$$

But binding energy (MeV) = mass defect (MeV/c^2)

From Table I the ratio of a.m.u. to MeV/c^2 can be deduced, for example using the $^{12}_6$C nucleus:

$$\text{Mass in a.m.u.} = \text{Mass in MeV}/c^2 \times \frac{12}{11178}$$

$$\text{mass defect } ^4_2\text{He} = 28.12 \text{ MeV}/c^2$$
$$= 28.12 \times \frac{12}{11178} \text{ a.m.u.}$$
$$\approx 3.0 \times 10^{-2} \text{ a.m.u.}$$

Question 12

A   fusion.

B   neither (but conventional methods using fossil fuels do).

C   fission (possible fusion processes are thought of as starting from non-radioactive materials).

D   fission (and fossil fuel methods; fusion processes are most likely to only need water!).

E   fission.

Acknowledgements

Grateful acknowledgement is made to the following source for the illustration used in this Unit:

Nuclear Power Group for Fig. 12.

The Open University

Science Foundation Course Unit 32

ELEMENTARY PARTICLES

Prepared by the Science Foundation Course Team

THE OPEN UNIVERSITY PRESS

Contents

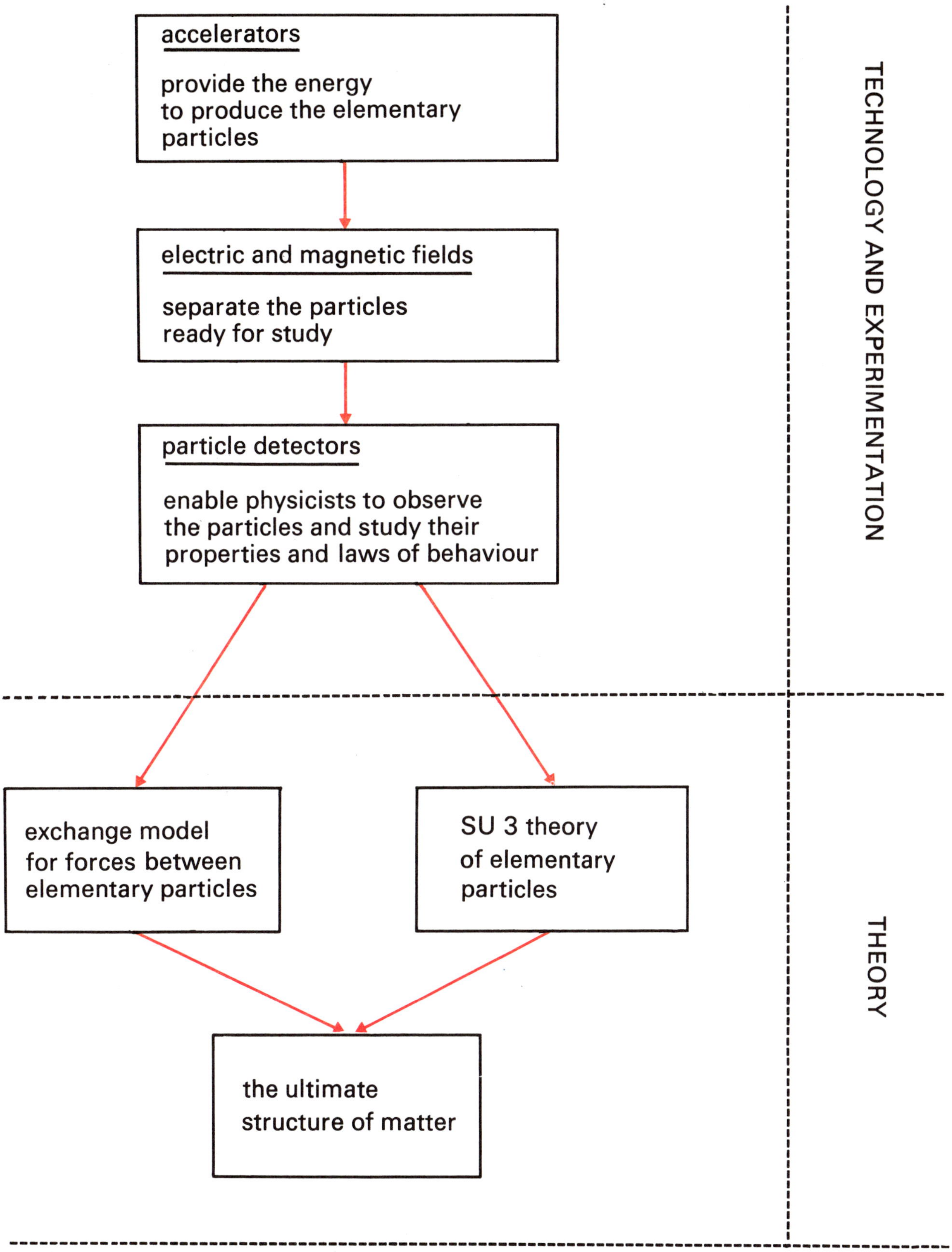

accelerators

provide the energy
to produce the elementary
particles

electric and magnetic fields

separate the particles
ready for study

particle detectors

enable physicists to observe
the particles and study their
properties and laws of behaviour

**exchange model
for forces between
elementary particles**

**SU 3 theory
of elementary
particles**

**the ultimate
structure of matter**

TECHNOLOGY AND EXPERIMENTATION

THEORY

Table A

List of Scientific Terms, Concepts and Principles used in Unit 32

Taken as pre-requisites			Introduced in this Unit			
1	**2**		**3**		**4**	
Assumed from general knowledge	Introduced in a previous Unit	Unit No.	Developed in this Unit	Page No.	Developed in a later Unit	Unit No.
	strong interaction	31	meson	12		
	electromagnetic forces and fields	4	elementary particle	13		
			pion	13		
	conservation laws:		linear accelerator	16		
	of energy	4	proton source	16		
	of momentum	3	drift tube	17		
	of electric charge	2	magnetic curvature of path proportional to particle's momentum	19		
	rest-mass energy	4				
	kinetic energy	4	proton synchrotron	19		
	uncertainty relation	29	bending magnet	24		
	electron-volt	4	electrostatic separator	25		
	nucleon	6	bubble chamber	26		
	voltage	4	bubble growth initiated by charged particles	26		
	frequency	2				
	ionization	6	superheating	26		
	temperature	5	dependence of boiling-point temperature on pressure	27		
	electron	2				
	photon	29	electron pair	32		
	Compton scattering of photons	29	positron	32		
			dependence of bubble density on particle velocity	34		
	periodic table of elements	7				
			conservation laws:			
			of baryons	35		
			of strangeness	36		
			SU3 classification of elementary particles	39		
			quarks	42		

Objectives

When you have completed the work of this Unit, you should be able to:

1 Define, or recognize adequate definitions of, or distinguish between true and false statements concerning each of the terms, concepts and principles in column 3 of Table A.

2 Distinguish between true and false statements concerning the principles of operation of a linear accelerator, a proton synchrotron and a bubble chamber.

3 Solve qualitative problems on the deflection of particles moving in electric and magnetic fields.

4 Make simple deductions concerning the momentum, electric charge, mass and velocity of particles causing tracks in a bubble chamber.

5 Apply the laws of conservation of strangeness, electric charge and baryons to various reactions (without it being necessary to remember the names and properties of the particles involved).

6 Given the values of the electric charge that can be carried by a particle, deduce the value of $(Q-\overline{Q})$ for the particle when it carries a particular charge.

7 Recognize the arrays representing SU3 groupings of 8 and 10 particles on a plot of strangeness versus $(Q-\overline{Q})$. Use such arrays to deduce the properties of missing particles.

Introduction

The study of elementary particles is the study of the structure of matter. It is appropriate that as our Foundation Course draws to its close we should turn to this most basic of modern scientific endeavours.

In this Unit you will learn how quantum behaviour, relativity and nuclear physics merge to create the fascinating world of microphysics. You will see for yourself violent collisions giving birth to particles possessing unfamiliar properties—properties with names like 'strangeness'. You will be led to question whether particles like protons and neutrons are really the fundamental basic building blocks of matter, or like atoms, merely composite structures built out of something yet more elementary.

The subject of elementary particles is often regarded as too 'theoretical', too 'abstract' to be really understood except by a handful of experts. Up to a point this is so. A full appreciation of the subject requires a grasp of advanced mathematics. Nevertheless we hope to show you that it is possible to go quite far into the subject before lack of mathematics becomes a bar to further progress. True, one needs a certain background knowledge of physics in order to make a start—but this background knowledge you have now acquired from the earlier parts of the Course.

By the end of the Unit you will have learnt that it requires only familiar things like hydrogen gas, electric and magnetic fields, and boiling liquids to produce and detect these particles. The technology involved is impressive, but the basic principles are easily understood. And as for this peculiar property called strangeness, by the end of the Unit not only will you understand why it is necessary to introduce such a concept, but you will also be able to deduce for yourself the strangeness values of the particles.

The Unit begins with an extended preamble. The purpose of this first section is to set the scene. It does so by describing the theory that started it all. The work of the Unit really begins with sections 2 and 3. These are devoted to a description of the remarkable tools of the high-energy physicist—particle accelerators and bubble chambers; it is with these that he produces new particles and studies their subsequent behaviour. In the fourth section, we give you a feel for the kind of discoveries that have been made. This is done in a rather novel way—in the form of a 'guided tour' of the world of elementary particles, as seen through bubble-chamber stereo-photographs. Finally we discuss where the subject might eventually lead.

Study Comment

Although you will be required to answer some questions which involve many particles and their properties, you are not required to remember or recognize the names or properties of any of them. These particular questions have been designed to test whether you know how to apply a few very simple rules of behaviour – the names and properties of the particles will be given to you as and when required.

32.1 Preamble

32.1.1 Yet another model for a force

You may remember from Unit 4 how difficult it was to 'explain' the concept of force. You learnt that it was something one *infers* from motion —when a particle moves in a certain way, it is said to be acted on by a force.

Various models of fields and forces were described—balls on rubber membranes and corks bobbing up and down on water. No one suggests these models are in any sense adequate, but from time to time they can be useful to illustrate a point. We now introduce you to yet another model. At first it will strike you as quite bizarre. Nevertheless, as you will soon learn, it has proved extraordinarily useful and successful.

Imagine you are an astronaut assigned to the task of exploring a new planet. You have arrived in the vicinity of the planet and have gone into orbit around it. As you look down, you see that large patches of the surface are green. On one of these green patches are two unidentified white objects—they are so far away they look like two white dots (Fig. 1 (a)). As you watch them you note that their separation, although it varies, never exceeds a certain maximum value. Other white dots come on to the green patch and move about singly—all except one which happens to pass close to the original two (Figs. 1 (b) and 1 (c)). When it comes within range of this maximum separation, its motion is arrested (Fig. 1 (d)). From that point onwards it stays close to the other two (Figs. 1 (e) and (f)).

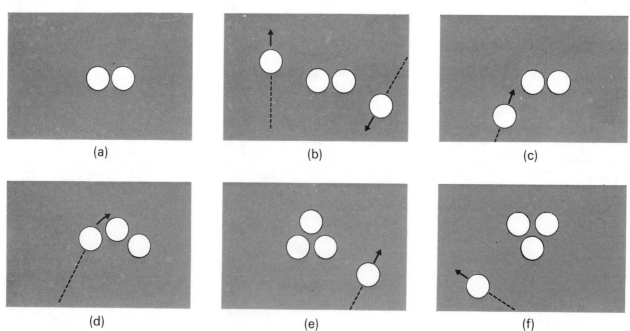

(a)   (b)   (c)

(d)   (e)   (f)

Figure 1   The behaviour of white objects on the surface of a hitherto unexplored planet.

What might you conclude from the behaviour of these dots? Is it reasonable to postulate a force of some kind acting between the white objects?

Yes, indeed. A force acting between the white objects would be a very reasonable way of describing their behaviour. The original two objects

were constrained to stay within a certain distance of each other because of a force of attraction between them. The newly arrived object happened to come within the range of this force and that is why it stuck to the other two. Indeed this force looks very much like the short-ranged nuclear force, or strong interaction, which we were discussing in Unit 31, and also like the force holding atoms together in solids and liquids (Unit 5).

In order to investigate the nature of the white objects and the force acting between them, you decide to descend and land on the green patch. What will you find—white balls with springs, or rubber cords stretched between them (Fig. 2)?

On landing, you find that the planet is exceedingly civilized—you have arrived in time for the start of a cricket match. Most of the players are just strolling around, but in one corner of the field there are three of them conscientiously practising their throwing and catching. Originally of course there were only two—as the third passed by he was thrown the ball and in that way joined them. You are now able to see that the maximum separation of the cricketers, i.e. the range of the supposed force, was nothing more mysterious than the maximum distance the ball could be thrown (Fig. 3, p. 12).

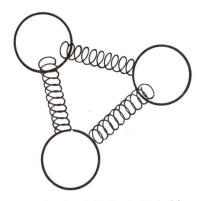

Figure 2 *Could this be the kind of force acting between the white objects?*

Suppose the cricketers decide to practice with a much heavier ball, what effect would this have on the characteristics of the 'force'?

The heavier the ball, the shorter would be the range of the force (Fig. 4). Thus it is concluded that:

a process involving the exchange of an intermediary object may give rise to effects similar to those produced by a force.

Forces that can be described in this way are not surprisingly called *exchange forces.*

exchange forces

32.1.2 The exchange model applied to the nuclear force

No one, of course, would seriously want to introduce the idea of cricketer-cricketer attractive forces. But the question does arise as to whether one ought to look at phenomena commonly called 'forces' and see if they can be described in terms of exchange processes. In particular, how about the nuclear force? This was the question posed by the Japanese physicist Yukawa in 1935.

What two important characteristics has the nuclear force?

As you learnt in Unit 31, the nuclear force is strong and has a range of only about 10^{-15} m. Is it possible to relate the mass of a supposed exchanged intermediary to this range, in the same way as the range of the hypothetical cricketer-cricketer interaction was governed by the mass of the ball?

Before answering that question, we must first consider a more fundamental one—how is it possible for the nucleons to exchange anything *at all*? If the two nucleons are at rest, their only energy is their rest-mass energy, so where do they find energy (i) to make the rest-mass of the intermediary object and (ii) to give it enough kinetic energy to go across the gap between them? The law of conservation of energy would appear to rule against it.

an application of Heisenberg's uncertainty relation

At this point, Heisenberg's uncertainty relation, involving uncertainty in energy, ΔE, and uncertainty in time, Δt, puts in an appearance:

$$\Delta E\, \Delta t \approx h/4\pi \dots\dots\dots\dots\dots (1)$$

In Unit 29, we were at pains to point out that the law of conservation of energy can be experimentally verified only to within the limits allowed by this relation. There is no evidence to support a view that the energy of a system must remain exactly constant at all times. If one finds it useful to postulate that the energy of the system might fluctuate by an amount ΔE for a length of time Δt, such a postulate could never run counter to any experimental evidence. (It might be a good idea to look back over section 29.4.1 of Unit 29 to refresh your memory as to the nature of this argument.)

Thus Heisenberg's relation makes it possible to suggest that the energy of the two-nucleon system could fluctuate by an amount ΔE for a period Δt. There could be a large fluctuation for a short time, or alternatively a smaller fluctuation for a longer time. The two nucleons could therefore 'borrow' energy ΔE on a long-term or short-term loan (naming Heisenberg as surety). This energy could then go towards creating an intermediary object which on being exchanged somehow produced the nuclear force. Once the exchange was completed, the process could be repeated again and again, so the force would apparently operate continuously.

We now return to the earlier question—can we from the range of the force discover the mass of the supposed intermediary particle? We can arrive at some idea of this quite quickly from the known separation of the nucleons, R, if we use equation 1. The intermediary cannot cross the gap between two nucleons in zero time. It follows from the work of Units 3 and 4 that no object can travel faster than c, the speed of light.*

Thus the minimum time, Δt, for which the loan of energy is required is the time taken to travel a distance R at a speed c:

$$\Delta t = R/c \quad \dots\dots\dots\dots\dots (2)$$

(If the speed is lower, the time interval must of course be longer.) Corresponding to this minimum time interval, there is a maximum energy that can be borrowed, and this is given by substituting Δt in equation 1:

$$\Delta E. \frac{R}{c} \approx h/4\pi$$

i.e.

$$\Delta E \approx \frac{hc}{4\pi R} \quad \dots\dots\dots\dots\dots (3)$$

Calculate ΔE in electron-volts, using the following values:

$h = 6.6 \times 10^{-34}$ J s
$c = 3 \times 10^{8}$ m s^{-1}
$R = 10^{-15}$ m
1 electron-volt $= 1.6 \times 10^{-19}$ joule

$$\Delta E \approx \frac{6.6 \times 10^{-34} \times 3 \times 10^{8}}{4\pi\; 10^{-15}}$$

$$\approx 16 \times 10^{-12} \text{ joule}$$

$$\approx \frac{16 \times 10^{-12}}{1.6 \times 10^{-19}} \text{ eV}$$

$$\approx 100 \times 10^{6} \text{ eV}$$

$$\approx 100 \text{ MeV}$$

(Remember you were introduced to electron-volts as a unit of energy in Unit 4.)

It should be remembered that the estimate of 100 MeV for ΔE is very crude. This is because equation 1 is only approximate.

---

* *You may remember that the momentum of an object is given by* $p = m_0\, v_{im}/(1 - v_{im}^2/c^2)^{\frac{1}{2}}$ *where* m_0 *is the rest-mass and* v_{im} *is the improper velocity (i.e. the velocity of the object as usually measured in the laboratory frame of reference). As the object travels faster, so its momentum increases. The expression shows that* p *becomes infinite as* v_{im} *approaches* c. *Thus no object can be accelerated to a velocity exceeding that of light.*

This calculation suggests that when nucleons are separated by a distance of 10^{-15} m they can exchange an intermediary with an energy up to 100 MeV. If they are closer than 10^{-15} m, then the time required for the exchange, Δt, is smaller and the uncertainty relation allows ΔE to be correspondingly higher. But what if the nucleons are much further apart than 10^{-15} m? As you already know—the force drops to zero. For our model of exchange forces, this implies that, for one reason or another, the exchange process ceases. One way of explaining this is to postulate that whereas the intermediary object can have energy 100 MeV or more, *it cannot have less than* 100 *MeV.*

An object can always be given *more* energy by giving it extra kinetic energy; but why do you think this particular object can never have *less* than 100 MeV?

Figure 3 In order to exchange a ball, the players must remain within a certain range of each other.

The indication is that *the intermediary particle has a rest-mass corresponding to an energy of* 100 *MeV.* In order for the particle to exist, it must have at least 100 MeV for its rest-mass. If the separation of the nucleons is sufficiently great that Δt in equation 1 requires ΔE to be less than 100 MeV, the particle cannot be exchanged and the force disappears.

This then was the substance of Yukawa's remarkable theory. He introduced an entirely new particle. It had a rest-mass energy of about 100 MeV, or if you like, a mass some 200 times that of the electron (the mass of an electron is equivalent in energy to about 0.5 MeV). Because this particle had a mass intermediate between that of the nucleons and electrons it was called a *meson* (from the Greek word 'mesos', meaning 'in the middle').

Figure 4 The heavier the ball, the closer the players must be.

32.1.3   The discovery of Yukawa's meson

Of course, the meson, as it has been presented to you so far, is a rather ethereal object! If it exists at all, it comes and goes, its fleeting existence being permitted only by the uncertainty relation.

meson

Its energy, ΔE, must be returned before the time Δt elapses. Like Cinderella, it is allowed out only on condition that it is back home on time! Thus far, the idea that the nuclear force is due to the exchange of mesons is nothing more than a theoretical model. But Yukawa went further.

Suppose the meson did not have to rely on the uncertainty relation for the energy it needed for its rest-mass—suppose there were some other supply of energy to hand. It would then no longer be necessary to return the energy after a limited time Δt, and the meson could take on a separate and more permanent existence. Such a situation exists when nucleons, instead of being stationary, are involved in a violent collision. The energy of the two-nucleon system is now greater by virtue of the kinetic energy possessed by the original particles. The nuclear force comes into operation at the moment of impact. The meson is now exchanged under conditions where a great deal of energy exists, and it can take some of the kinetic energy of the two nucleons for its rest-mass energy. The law of conservation of energy can be rigorously obeyed in as much as the energies before and after the collision can be the same, even though a meson has now been brought into existence:

$$Mc^2 + Mc^2 + T_1 + T_2 = Mc^2 + Mc^2 + mc^2 + T'_1 + T'_2 + T'_m \cdots \cdots \quad (4)$$
$$\text{(before the collision)} \qquad\qquad \text{(after the collision)}$$

where M is the mass of each nucleon and Mc^2 is the equivalent rest-mass energy (see Unit 4); T_1 and T_2 are the initial kinetic energies of the nucleons and T'_1 and T'_2 are the final kinetic energies; mc^2 is the rest-mass energy of the meson and T'_m its kinetic energy.

A proton of kinetic energy 1 000 MeV strikes a stationary nucleon. After the collision, the kinetic energies of the two nucleons are 500 and 200 MeV. What is the kinetic energy of the meson produced in the collision? (Assume it has a mass, m, such that $mc^2 = 100$ MeV.)

Therefore, on the basis of Yukawa's model, it was possible to predict that in a violent nuclear collision, in which sufficient energy was available to create the necessary rest-mass, a physically detectable meson might be produced (see Fig. 5).

$Mc^2 + Mc^2$ cancels out on both sides of the equation. $mc^2 = 100$ MeV, $T_1 = 1\,000$ MeV, $T_2 = 0$, $T'_1 = 500$ MeV, $T'_2 = 200$ MeV

$\therefore\ 1\,000 + 0 = 100 + 500 + 200 + T'_m$

$\therefore\ T'_m = 200$ MeV

nucleons

meson   nucleons

Figure 5   *In high-energy nuclear collisions, mesons may be produced.*

This prediction was triumphantly vindicated in 1947 by Powell, Occhialini and Lattes with the discovery of Yukawa's meson—now called the π-meson, or *pion* for short. Its mass of 273 times the electron mass (or in energy units, 139 MeV) was very close to that predicted by Yukawa. The way to a clear understanding of nuclear forces seemed at that time straight-forward. All one had to do was to produce pions, study the new particles' properties and observe how they were scattered and absorbed by nucleons.

the discovery of Yukawa's meson—the pion

This mood of optimism was soon shattered. In the same year that the pion was discovered, *another* and heavier particle was found among the debris of a high-energy nuclear collision. Then more and more appeared. Today the nuclear scientist is confronted by a bewildering array of about 200 so-called 'elementary' particles. (The words 'elementary' and 'fundamental' are used interchangeably in connection with these particles. An *elementary particle* is one that cannot be described as a composite structure of more basic components. Incidentally, the choice of the word 'elementary' might suggest that the properties of these particles are simple—nothing however could be further from the truth!)

elementary particle

The model whereby nucleons exchange a pion tells only part of the story. It may be adequate for nucleons separated by 10^{-15} m, but when nucleons are closer it appears that much heavier objects can be exchanged as well.

One of the new particles is found to have a rest-mass energy of 550 MeV. If pion exchange extends to 10^{-15} m, how close do the nucleons have to be in order to be able to start exchanging this other particle?

A proper understanding of the nuclear force then seems to entail an understanding of the behaviour of all these new elementary particles. This would be impossibly complicated if it were not for the existence of certain simplifying rules and patterns of behaviour; we shall discuss these towards the end of the Unit, having first described how elementary particles are produced and detected in the laboratory.

The mass of this other particle is about 4 times greater than that of the pion, so the separation of the nucleons must be correspondingly smaller, i.e. 0.2 to 0.3×10^{-15} m.

32.1.4 Summary of the preamble

A new model for describing forces has been introduced. According to this model, a force acting between two or more objects may be represented by the exchange of some intermediary between them.

This model has been applied to the nuclear force. From the uncertainty relation and the known range of the force, the mass of the intermediary, called a pion, could be determined; it is 273 times the mass of the electron (i.e. has a rest-mass energy of 139 MeV).

When sufficient energy is available, as it is in a very violent nuclear collision, some of it may transform into the rest-mass energy of the intermediary pion which can thereby take on a real and separate existence.

Other particles have also been discovered. The exchange model of the force must therefore be developed further to allow the nucleon to exchange additional heavier particles during close approaches.

The study of the nuclear force therefore becomes the study of elementary particles.

the study of the nuclear force becomes the study of elementary particles

Figure 6   An aerial view of CERN, Geneva.   [Photo: CERN]

32.2 Particle Accelerators

32.2.1 The need for sophisticated equipment

There are two basic experimental requirements before elementary particles can be produced and their behaviour studied.

What do you think these basic requirements might be?

In the first place the physicist needs a source of highly energetic nuclear particles, such as protons, with which to bombard other protons and produce the new particles. The energy must exceed several thousand MeV to be of interest—this figure being set by the masses of the heaviest particles he wishes to produce. (To appreciate how large this energy is, you should note that the typical kinetic energy of an atom moving about at room temperature is only a few hundredths of an electron-volt.)

Secondly, it is necessary to have devices capable of detecting particles as small as a nucleus, moving with speeds approaching the maximum value, i.e. the speed of light. This means extending the senses to the very frontiers of small-scale and rapidly changing observations (Unit 2).

In response to the demands of the high-energy physicist (i.e. a physicist who studies elementary particles), technologists have produced some of the world's most remarkable machines—the particle accelerators. These are now so complex and costly that only the U.S.A., Russia, and lately Japan, can afford to finance them alone; the European nations have to pool their efforts and share a common facility in Geneva, Switzerland at CERN (Conseil Européen de la Recherche Nucléaire).

In this section we give a description of the principles of these machines. In this Unit's television programme we shall take you on a visit to CERN (Fig. 6) to show you the actual equipment.

32.2.2 The protons start their journey

It is not difficult to get an accelerated beam of protons of low energy (up to 1 or 2 MeV).

Think back over the Course. Can you remember a previous occasion when an accelerated beam of positive ions was obtained?

In Unit 6 it was described how, in the mass spectrometer, an electric discharge passing through a gas, ionized the gas atoms. The positively charged ions could then be accelerated by subjecting them to an electric field. Exactly the same idea is used again here. In the *proton source*, hydrogen gas is ionized so as to give free protons. The principle is illustrated in Figure 7. These are then subjected to an accelerating voltage. An energy of say 1 000 MeV would be achieved by accelerating the protons through a voltage difference of . . .

how many volts?

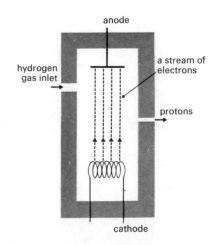

Figure 7 A source of protons. Hydrogen gas enters and is ionized in an electrical discharge.

1 000 MV. (Refer if necessary to Unit 4.) This is the first major problem—
it is quite impossible in the laboratory to sustain an electric field capable
of giving a voltage difference of anything like this amount. In practice it
has proved impossible to exceed about 10 MV.

Can you guess why?

Long before one gets to the kind of voltage that interests a high-energy
physicist—lightning strikes! With a deafening bang the high-voltage
electrodes discharge, either to each other or to the surroundings (Fig. 8).

'Brute force' is therefore no answer. Somehow the protons must be coaxed
to go at a speed corresponding to an enormous voltage, *without the use of
such a large voltage.*

a barrier to further acceleration

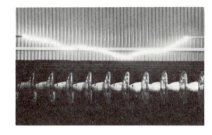

*Figure 8   Electrical discharges like this
one set a limit on the voltage differences
that can be realized in practice.*

32.2.3   The linear accelerator

This can be achieved in a *linear accelerator*, a name often contracted to
linac. It consists of a series of cylindrical metal tubes arranged along a
common axis. They are connected to a voltage supply as shown in
Figure 9.

linear accelerator

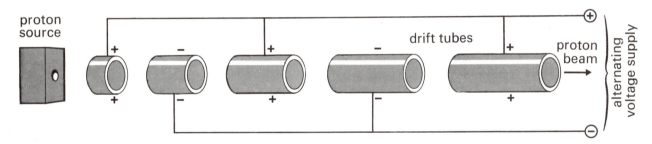

*Figure 9   A schematic diagram of a linear accelerator. The protons are repeatedly
accelerated in small stages as they cross the gaps between the drift tubes.*

The whole system is evacuated so that the protons from the source can
pass down the axis of the tubes and suffer little scattering from air molecules. To see how the protons are affected by the voltages on the tubes,
take a look at Figure 10 where the configuration of the electric fields is
shown.

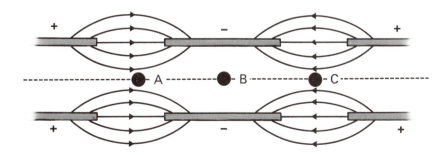

Figure 10   *The electric field configuration in a linear accelerator.*

**Assuming the proton moves from left to right along the axis in Figure 10,
what change if any do you expect to its velocity in each of the regions labelled
A, B and C?**

16

As the proton crosses the gap between the tubes at A, the direction of the electric field is such as to increase its velocity. At C the field is in the opposite direction and in this region the proton would tend to slow down. When the proton is well inside a tube, for example at position B, there is no field so the proton 'drifts' for a while at constant velocity—for this reason the tubes are called *drift tubes*.

drift tubes

What is the net effect on the proton's velocity in passing from A to C?

Nil! The trouble of course is that the slowing down in region C exactly compensates for the acceleration at A. Fortunately, it is possible to do something about this. While the proton is drifting in the field-free interior of a tube, the polarity of the voltage supply (see the right-hand side of Fig. 9) can be switched—the tubes that were positive now become negative and vice versa. This can be achieved without in any way affecting the proton's motion.

If this change in polarity occurs while a proton is at position B in Figure 10 what will happen to it when it emerges from the tube at C?

The change in polarity converts retarding fields into accelerating ones and vice versa. *A proton originally accelerated across the gap at A, finds itself again accelerated at the next gap C. (See Fig. 11.)*

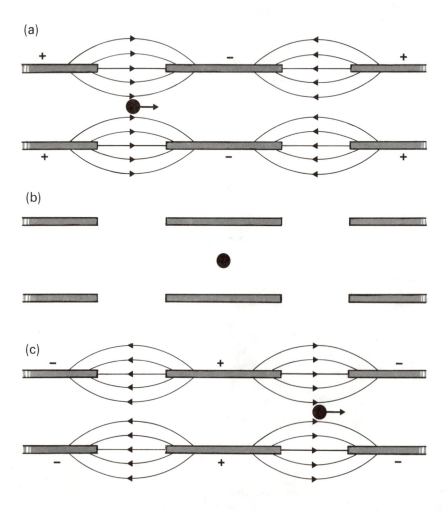

(a)

(b)

(c)

Figure 11 Successive field configurations as a proton moves from one gap to the next.

If the voltage between the tubes is 50 kV, what is the net gain in energy of the proton after it has crossed two gaps?

It gains 50 keV at each gap, so the total is 2×50 keV = 100 keV.

The polarity of the voltage can be switched repeatedly by an alternating voltage supply so that the process can be continued. The proton gains an increment of energy corresponding to the voltage difference between the tubes each time it crosses a gap. If the voltage difference is V volts, the energy of a proton that has crossed n gaps is nV electron volts.

Take another look at Figure 9. What difference do you notice in successive drift tubes? Why should there be this difference?

If the voltage polarity changes at regular intervals (i.e. at constant frequency), the drift tubes must become progressively longer to allow for the proton's increasing speed—otherwise they would get out of step with the accelerating field across the gaps.

(In modern linear accelerators, the voltages are not applied in quite the way we have described here. The tubes are mounted along the axis of a hollow cylindrical cavity (shown opened in Figure 12), and currents of a suitable distribution are set up in the walls of the cavity to produce the fields. Such an arrangement leads to an improved electrical efficiency. The details of this refinement need not concern you, however, because the principle remains the same—the protons are progressively accelerated at each gap.)

Figure 12 *Part of a linear accelerator opened up to show the drift tubes. The proton source is at the far end.*

The way is now open to enormous energies—at least from the technical point of view. All we have to do is to have lots and lots of drift tubes. But a problem of another kind arises at this point.

The linear accelerator in Figure 12 delivers protons of energy 50 MeV. The CERN machine was to reach an ultimate energy of 28 GeV, i.e. 28 000 MeV (1 Gev ≡ 10³ MeV ≡ 10⁹ eV). This could have been reached by simply extending the linear accelerator further, but what kind of problem do you think would have been encountered?

Certainly one important problem would have been that of money—how many miles of accelerator can a government be persuaded to finance! (As a matter of interest, the longest linear accelerator in the world is a machine for accelerating electrons to 20 GeV at Stanford, California; it is two miles long.)

a further barrier to the achievement of greater energies

32.2.4   The proton synchrotron

The synchrotron introduces another ingenious idea in accelerator design, one that allows high energies to be achieved more economically than by extending a linear machine. In the synchrotron, the protons are made to travel many times around a circle of constant radius. (They must, of course, still be travelling in a vacuum.) Cavities are placed at various points around the circumference and these provide electric fields for acceleration. By repeatedly following the same path, the protons are accelerated *many times by the same cavities* (see Fig. 13).

proton synchrotron

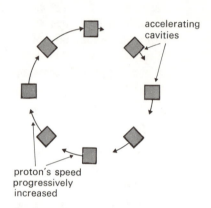

In order to keep the protons on the circular path, a centripetal force must be provided at right-angles to their motion. You will remember that in a mass spectrometer (Unit 6) the heavy ions were deflected by a magnetic field. The same also holds in a proton synchrotron; a vertical magnetic field steers the protons along their horizontal circular path (see Figs. 14 and 15).

Figure 13   In a proton synchrotron, the protons are repeatedly accelerated by the same cavities.

Just as in the mass spectrometer, *the deflection produced by a magnetic field of a given strength is such that the radius of curvature of the particle's path is proportional to the particle's momentum.* (This is an important relationship which you should remember.)

If the protons are to be accelerated, *and yet remain on a path of constant radius*, what must happen to the magnetic field?

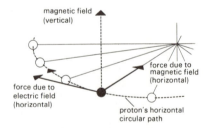

In order to keep the accelerating protons moving along the same path in the synchrotron, the strength of the magnetic field must be progressively increased. This is achieved by progressively increasing the electric current in the windings of the electromagnets during the acceleration period.

Figure 14   In a synchrotron, the proton is accelerated by electric forces acting tangentially to its circular path. Meanwhile the magnetic field exerts the necessary restraining force directed towards the centre of the circle.

As the proton moves faster, the time it takes to arrive at each successive accelerating cavity becomes shorter and shorter. In the linear accelerator it is possible to make the distance between the successive accelerating gaps longer so that the proton does not get out of step with the fixed frequency accelerating voltage. This is not possible in the synchrotron because the proton traverses the *same* cavities many times and on each occasion at a different speed.

How do you suggest the proton's arrival at the accelerating cavities could be kept in step with the alternating voltage?

The answer is to abandon the idea of a *fixed* frequency alternating voltage. Instead the frequency is progressively increased so as to keep exactly in step with the arrival of the protons at the cavities; in short, they are synchronized—hence the name of the machine. Does this mean that we

Figure 15   An analogy for a synchrotron.

are now able to reach any energy we like by simply sending the protons round and round repeatedly? Unfortunately, no. As the protons accelerate they require an ever stronger magnetic field to hold them on course. Finally it becomes impracticable to increase the field any further and no more acceleration can take place—the protons have reached the maximum energy appropriate to the particular accelerator.

Having reached the maximum energy the protons are deflected out of their circular path by the sudden introduction of an additional local magnetic field. They are directed on to a target consisting of a lump of metal. As the protons crash into the nuclei of the atoms of the target material, the new elementary particles are created.

32.2.5 The CERN 28 GeV accelerator—a summary of accelerator techniques

To reach the highest energies, the three accelerating techniques (high voltage, linear accelerator and synchrotron) are combined into one machine, and the protons are accelerated in three stages. We now describe one of these machines—the one at CERN. In so doing we can provide you with a short summary of the previous sections. Incidentally, you are not required to remember any of the numbers quoted in the summary that follows; they are there merely to give you a feel for the orders of magnitude involved.

(i)   The protons to be accelerated are produced through the ionization of hydrogen gas in an electrical discharge.

(ii)   At the start of the acceleration cycle, the protons are given an initial acceleration by a large voltage difference of 550 kV (see Fig. 16).

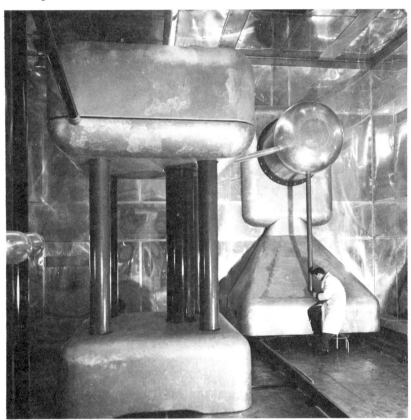

Figure 16   *The protons start their journey at CERN in this metal-cladded room, called a Faraday cage. The source is situated in the horizontal cylinder on the left, above the technician. When in operation the source is put at a voltage of 550 000 V with respect to the walls of the cage. In the foreground, mounted on ebonite pillars, is a platform housing the control equipment for the source. The protons are fired from the source into the linear accelerator which is situated beyond the far wall.*   [Photo: CERN]

(iii) They pass into a linear accelerator. An alternating electric field is set up and this acts upon the protons only when it is in a sense such as to cause acceleration. At other times, the protons are shielded from the electric field by drift tubes. The particles are accelerated in this way to an energy of 50 MeV.

(iv) The protons are injected from the linear accelerator into the synchrotron (see Fig. 17). They are steered in a horizontal circular path of diameter 200 m by a magnetic field produced by electro-magnets (Fig. 18). They are accelerated at fourteen points around the circumference of the circle by electric fields. A drawing of a section of the synchrotron is shown in Figure 19.

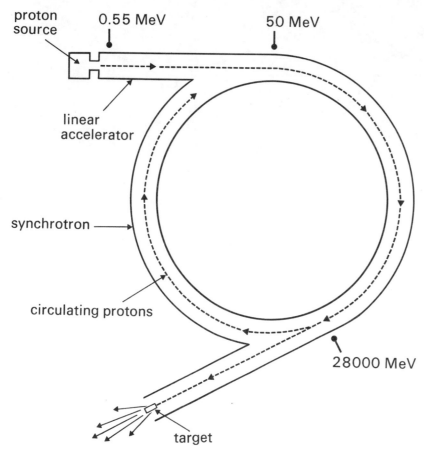

Figure 17   *The CERN 28 GeV accelerator. It combines three types of acceleration: high voltage, linear accelerator, and synchrotron. On reaching the maximum energy the protons are ejected and strike a target, producing secondary particles.*

Figure 18   *One of the 100 32-ton electromagnets used for steering protons around the 200 m diameter ring of the CERN synchrotron. Passing along the centre of the magnet you can see the evacuated tube within which the protons travel.*   [Photo: CERN]

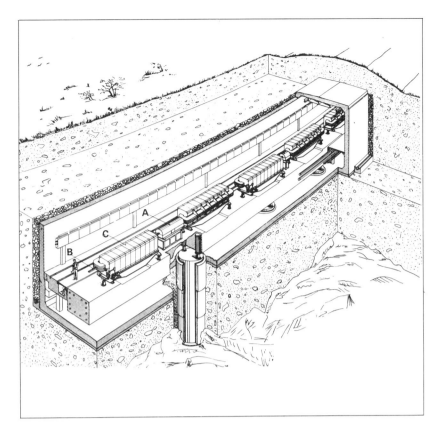

Figure 19   *A section of the synchrotron at CERN. B is the evacuated tube through which the protons pass. C is one of the fourteen accelerating cavities distributed around the ring. A are electromagnets. The machine is housed in an underground tunnel to make it easier to confine the stray radiation produced during its operation. The alignment of the magnets has to be maintained to a fraction of a millimetre. For this reason they rest on a concrete base which is itself elastically supported by massive columns standing on firm bed-rock.*

The magnetic field is progressively increased in order to keep the protons on course, and the frequency of the alternating electric field is increased to keep in step with the accelerating protons. The acceleration cycle is completed in a total of about two seconds. In this time the protons have been round the synchrotron 480 000 times (a distance equivalent to going several times round the Earth). On reaching their maximum energy of 28 GeV the protons are ejected from the machine.

(v) Another batch of protons is admitted and the process repeated.

(vi) The net result is that about 10^{12} protons, each of energy 28 GeV, are emitted in bursts at two-second intervals. These are directed on to a metal target.

Try SAQs 1 to 7.

32.3 Experiments with Particles

32.3.1 Sorting out the particles

Once the elementary particles have been produced in the target, the next problem is to investigate their properties. This is best done by separating out one particular type of particle at a time from all the others. Having studied that particular particle, one can then systematically move on to consider the next. In this Unit's main home experiment, you will be concentrating on a particular investigation with certain limited objectives in mind. For this to be possible, pions had to be separated out from all the other particles and nuclear debris coming away from the target. But how was this done? How can one get rid of all the other particles, and leave only those with the desired mass, momentum and electric charge, all moving in the desired direction?

A start may be made by isolating those particles moving in one chosen direction.

Can you guess how this might be achieved?

This can be done with the help of a heavy shielding wall in which there is simply a small hole to let the particles through (see Fig. 20).

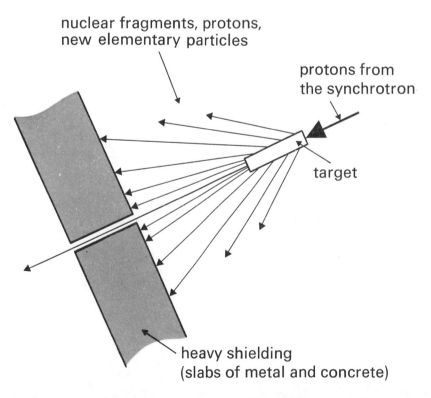

Figure 20 A slit in a heavy shielding separates out those particles moving in one given direction. All other particles slow down and undergo interactions in the material of the wall.

23

Which of the following characteristics do you expect for the particles emerging from the hole:

(i) same or different masses?

(ii) same or different momenta?

(iii) same or different electric charges?

(iv) same or different directions?

In the next step, one can separate out the particles according to their momenta.

(i) different
(ii) different
(iii) different
(iv) same

How could this be done?

This is achieved with a magnetic field. You will remember in the discussion of the synchrotron we mentioned that the radius of the curvature given to the path of a moving charged particle by a magnetic field was proportional to the particle's momentum. A vertical magnetic field therefore fans the particles out in a horizontal plane, as in Figure 21. The magnetic field also provides an additional bonus—it separates out particles according to their electric charge. The particles generally carry either one unit of positive charge (like the proton) or one unit of negative charge (like the electron) or are electrically neutral. If the positively charged particles are deflected to the right, then the negatively charged ones go to the left, and the neutral ones straight on.

A slit in a second heavy shielding wall is introduced and this gives rise to further selection (Fig. 22).

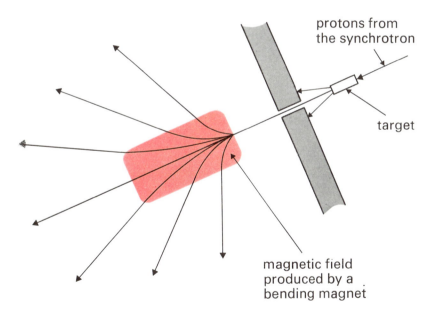

protons from
the synchrotron

target

magnetic field
produced by a
bending magnet

Figure 21   A magnetic field, on the far side of the shielding wall from the target, deflects the particles according to momentum and electric charge. We show a selection of trajectories. This deflection is achieved with an electromagnet called a bending magnet.

bending magnet

Which of the following characteristics do you expect for the particles emerging from the second slit:

(i) same or different masses?

(ii) same or different momenta?

(iii) same or different electric charges?

(iv) same or different directions?

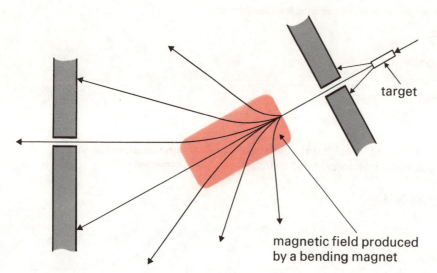

target

magnetic field produced
by a bending magnet

Figure 22   A second slit allows further selection of the particles to be made.

It is now necessary to separate the particles according to their mass. This can be done by a method analogous to that used in a mass spectrometer. In that instrument, as you may recall from Unit 6, ions of unequal mass were separated by first subjecting them to an electric field and then to a magnetic field. The particles from the synchrotron have already passed through a magnetic field; it remains for them now to be subjected to an electric field.

The particles are passed through the electric field of an *electrostatic separator*. As they move between two charged parallel metal plates, they are deflected sideways by an electric field acting perpendicular to the direction of motion of the particles. The particles all experience the same force because they all have the same electric charge.

electrostatic separator

Does this mean they are deflected in the same way?

They are not deflected in the same way, because particles of the same momentum but different mass move with different velocities and so spend different times within the region of the influence of the electric field. The result is that the particles are separated according to their mass (see Fig. 23).

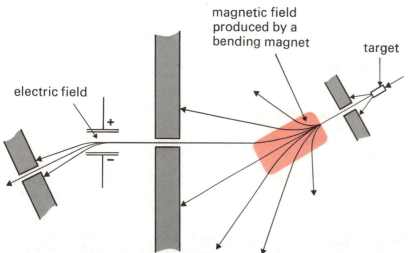

magnetic field
produced by a
bending magnet

target

electric field

Figure 23   The final stage in the sorting out procedure is performed by an electrostatic separator. In this instrument the particles pass through an intense electric field, and are deviated once more.

Which of the following characteristics do particles emerging from the third slit in Figure 23 possess:

(i) same or different masses?

(ii) same or different momenta?

(iii) same or different electric charges?

(iv) same or different directions?

The desired goal has now been reached—the final beam consists of particles of the same mass, momentum, electric charge, and direction.

final selection of particles completed

You might at this point like to try SAQs 8 and 9.

32.3.2   The bubble chamber

If the properties of elementary particles are to be investigated, not only do they have to be produced and separated, but their subsequent behaviour must be studied. The problem now becomes one of observation. How is it possible to see a particle as small as a nucleus travelling with speeds approaching that of light?

The high-energy physicist has in fact several tools to choose from. You were introduced to two of them in the TV programme of Unit 2; one was a cloud chamber and the other a scintillation counter. In Unit 31 you came across a third—a Geiger counter.

In order to render visible something as minutely small as a sub-nuclear particle, these and other detecting instruments must exploit some kind of instability. The general idea is rather similar to what happens in a forest fire. It starts from something very small like a lighted cigarette end. This in itself, of course, does not provide much of the energy of the conflagration—that is to be found in the dried up leaves and trees. The significance of the cigarette end is that it gets the process started; without it the fire would not have begun.

So it is with particle detectors; each of them depends upon some kind of instability. The effects produced directly by the sub-nuclear particle are tiny, but they are sufficient to initiate a large-scale process.

As we said, there are several types of instrument in use. Each has its own characteristic advantages and disadvantages. We shall concentrate on just one of these techniques—one of the more important called the *bubble chamber*.

bubble chamber

The type of instability used here is that of *superheating*. A liquid is said to be superheated when its temperature is above boiling point and a small disturbance of some kind will start the liquid boiling.

superheating

Try this experiment.
Take the boiling tube supplied in the Home Kit (it is the tube packed separately from the other test tubes), put some water in it, and heat it over the burner. Eventually the water will start to boil. Look carefully at the manner in which the water boils. Do the bubbles start to grow from points anywhere within the volume of water, or do they start growing in only certain preferred places?

The bubbles start growing where the liquid comes into contact with the walls of the tube rather than in the interior of the liquid. Moreover if you look carefully you will see that certain specific points on the walls of the tube

In order for a bubble to grow it must generally have some 'centre' from which to start. Usually this consists of some irregularity in the containing vessel—a sharp corner where the sides join or perhaps an imperfection in the surface. *Ionized atoms can also provide suitable centres for bubble growth.* As a charged particle moves through a liquid, its electric charge

ionized atoms can provide suitable centres for bubble growth

26

interacts with that of the electrons belonging to the atoms of the liquid lying in its path, and some of the atoms are ionized. The energy deposited in this process gives rise to localized heating. So the charged particle in effect leaves a trail of 'hot spots'. If the liquid is already superheated, the extra high temperature along the particle's path can be sufficient to cause bubbles to start to grow.

In a bubble chamber, the liquid is superheated, not by increasing its temperature, as you did with the water in the boiling tube—that would be rather slow—but by suddenly decreasing the pressure acting on it. *The temperature at which a liquid boils depends on the pressure.* If the pressure is suddenly lowered, the boiling-point temperature is lowered to the value appropriate to the new reduced pressure. In a bubble chamber, it is arranged that the temperature of the liquid (which remains essentially constant) is below the higher boiling point corresponding to the initial higher pressure, but above the lower boiling point corresponding to the lower final pressure. In this way the liquid can be *abruptly* thrown into a superheated state.

seem to be preferred over others; they give rise to an almost continuous stream of bubbles. These are places where there might be a speck of dust or an imperfection in the glass surface. If an object with sharp edges or points (a pin, or match stick, for example) is dropped into the water, then the bubbles will tend to grow from the object.

Now try Home Experiment No. 1, which demonstrates this method of producing a superheated state. In it you will use cold water to boil hot water!

A bubble chamber consists of a hollow metal chamber with glass windows in the side for viewing the bubbles. It contains a transparent liquid (usually liquid hydrogen). Bubbles in the liquid can be illuminated by light flashes and photographed with a stereocamera (Fig. 24). The sequence of operations in a bubble chamber can be summarized as follows:

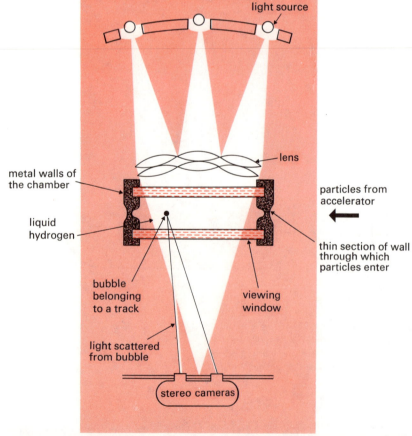

Figure 24   A simplified plan view of the optical system of the CERN 2-metre hydrogen bubble chamber. Light from three sources is concentrated by lenses in such a way that normally it misses the cameras. Bubbles interrupt this light and scatter it to the cameras, thus giving an image.

(i) A short while (about 10 milliseconds) before the particles are due to arrive from the accelerator, the pressure on the liquid is released by raising a piston (Fig. 25 (a)).

(ii) The particles pass through the thin metal wall of the chamber and into the liquid. They ionize the atoms of the liquid in their path and this process gives rise to local 'hot spots' (Fig. 25 (b)).

(iii) Bubbles start to grow from the 'hot spots' and continue to do so until they reach a visible size, whereupon the lights are flashed and stereophotographs of the bubble tracks are taken (Fig. 25 (c)).

(iv) The pressure is reapplied so as to collapse the bubbles (Fig. 25 (d)).

(v) The chamber is now ready to start another cycle.

You might at this point like to try SAQs 10 to 12.

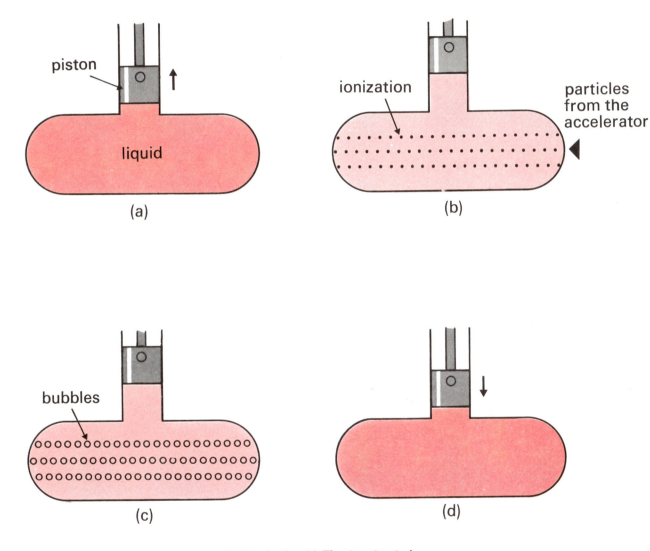

Figure 25 The sequence of operations in a bubble chamber. (a) The piston is raised, so releasing the pressure on the liquid and making it superheated. (b) Particles from the accelerator pass into the chamber and ions are formed. (c) In a few milliseconds the bubbles grow to a visible size. Stereophotographs are now taken by flashing the lights. (d) The piston is lowered to its original position, so restoring the initial pressure. The bubbles collapse.

32.4 A Look at Elementary Particles

In section 1, you were introduced to the ideas of Yukawa. At that time they no doubt appeared rather strange and abstract to you—far too 'theoretical' to have much to do with the 'real' world. We now take up the story again, but this time from a different point of view. Now we can show you the fascinating world of elementary particles as seen by the experimental physicist—as it is revealed in his detecting instruments and in particular in the bubble chamber.

For this purpose you will need the View-master supplied in the Home Experiment Kit; you will also need the reel of stereophotographs of a hydrogen bubble chamber entitled 'Reel 1, Detection of Nuclear Particles'. Insert the reel in the View-master such that the arrow pointing to the letter 'V' is at the top (Fig. 26 (a)). Depress the lever on the right-hand side of the viewer; this should bring up the first picture (Fig. 26 (b)).

Figure 26   *How to load your View-master with the reel of bubble-chamber photographs.*

The number of the picture and its title appear in the gap under the name 'VIEW-MASTER'. To see the pictures, you should hold the viewer up to a window or other reasonably strong source of light (Fig. 27).

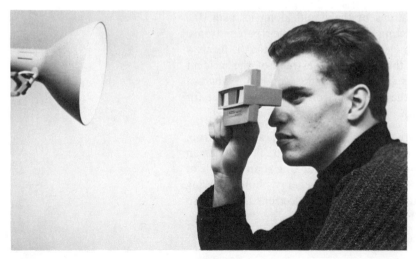

Figure 27   *Hold the View-master up to a good source of light.*

(Some people have a little difficulty at first in fusing the two pictures into a three-dimensional image—so be prepared to persist at it for several minutes. If you still have difficulty, it is conceivable that your eyes may be set unusually close together or wide apart. If this is the case you will not be able to see through both lenses at the same time. This will deprive you of the three-dimensional effect; but you will still be able to appreciate most of what is going on if you look through each lens separately.)

To move from one picture to the next, depress the lever on the right-hand side of the viewer.

32.4.1 Some general features of bubble-chamber photographs

Most people find bubble-chamber photographs very confusing at first. So before moving on to the more interesting events, let us just point out a few common features to help you feel more at home. Take a look at the first stereopicture.

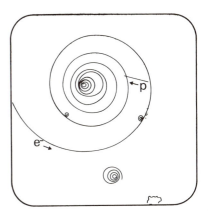

Figure 28 *Diagram for stereophotograph No. 1: spiralling electron.*

No. 1 Spiralling electron and stray proton

(See Fig. 28.)

The picture is dominated by an electron that enters the chamber from the left-hand side and spirals to rest. The curvature is produced by a magnetic field acting over the whole area of the chamber. The lines of magnetic induction are directed along the line of view. *Nearly all bubble chambers are provided with such a magnetic field.*

What information can be gained about the particle from the curvature of its track?

A measurement of the curvature yields an estimate of the particle's momentum.

Why does the radius of curvature of the electron track become progressively smaller?

As a charged particle moves through the liquid it continually loses energy in ionizing the atoms lying in its path. Its momentum decreases and so also must its radius of curvature in the magnetic field. This continues until the particle either leaves the chamber or comes to rest in the liquid.

In the process of ionizing the atoms of the liquid along its path, the moving particle is, of course, ejecting electrons from their parent atoms. Sometimes an atomic electron is ejected so energetically that it forms a bubble track of its own. When this happens, the track of the incident particle is seen to sprout a tiny spiral from its side. You can see two of these small spirals attached to the main electron track. Watch out for further examples in later pictures.

Towards the bottom of the picture you can see another spiralling electron. Unlike those you have considered so far, this electron track starts off in the liquid for no apparent reason. In fact you are looking at an example of the Compton scattering effect (Unit 29). A high-energy photon moving through the liquid collided with one of the atomic electrons and hit it so hard that the electron formed the track you see.

But where is the track of the photon?

Photons leave no tracks. In common with the other uncharged particles, they can leave no trail of ionization behind them, and therefore no centres upon which bubbles could form.

The interpretation of bubble-chamber photographs therefore involves making allowances for unseen neutral particles.

neutral particles leave no tracks

Just above and to the right of the middle of the picture is a short straight track starting and ending in the liquid.

How does its momentum compare with that of the electrons you have been looking at?

The straightness of the track indicates that the particle causing it had a much higher momentum than that of the electrons. In fact, the particle was a proton. Initially it was the nucleus of one of the hydrogen atoms of the liquid. It was probably struck by a fast neutron entering the chamber (the neutron leaving no track because, of course, it is uncharged).

Finally note the plume of bubbles growing at the very bottom of the picture (looking rather like the mushroom one associates with a nuclear bomb explosion). This is an example of bubble growth from some physical feature of the walls of the containing vessel. You will see other examples of this spurious boiling on some of the later photographs.

Now move to the next stereophotograph.

No. 2. Proton Scattering

(See Fig. 29)

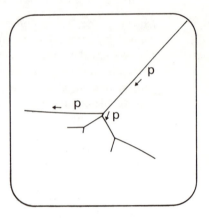

Figure 29 *Diagram for stereophotograph No. 2: proton scattering.*

A proton enters from the top right-hand corner leaving behind a continuous trail of ionized atoms. After a while it scores a direct hit on the nucleus of a hydrogen atom belonging to the liquid (by this we mean it comes within range of the strong interaction of the nucleus), and is deflected. From the collision come the two protons each leaving a track. This collision is followed by two further collisions.

interactions with the nuclei in the liquid

This photograph illustrates two distinct types of interaction that take place when a charged particle moves through matter:

(i) the frequent collisions with atomic electrons giving ionization and hence bubbles; and

(ii) the much less frequent—but more interesting—nuclear collisions.

As you have no doubt already noticed, there are also in the picture several low-energy electron tracks produced by the Compton scattering of photons.

There are about 10 particles entering the chamber from the top of the picture. These are particles associated with the beam from the accelerator. Are they positively or negatively charged?

Clue. **Which way do the electron tracks bend?**

The track of the positively-charged particle entering from the upper right-hand corner of the picture is steadily deflected to the right by the magnetic field, whereas those of the negatively-charged Compton electrons curve to the left. The particles entering the chamber from the top of the picture move to the left, and so they must be negatively charged.

32.4.2   The conversion of electromagnetic and kinetic energy into rest-mass energy

No. 3 Electron pair and triplet

(See Fig. 30)

In this picture, you can see one of the most important effects described by modern physics—the conversion of energy in the form of electromagnetic energy into energy in the form of rest-mass energy. It is sometimes referred to as the 'creation' of particles. This, however, is a little misleading because, of course, particles with rest-mass cannot be 'created' out of nothing. Their rest-mass is a form of 'potential energy' (Unit 4). So if a particle is to be produced, some already-existing energy in another form has to be transformed into the necessary rest-mass energy. Strictly speaking, we are concerned with *transformation* rather than creation.

A high-energy photon when it passes into an electric field can be transformed into a pair of electrons. One is negatively charged and is like any ordinary electron, the other has a positive charge and is either called a positive electron or a *positron*.*

Note that the law of conservation of electric charge is satisfied, because the net charge of the final particles is zero, like that of the original photon. The V-shaped pair of tracks towards the upper left-hand corner of the picture is an example of such an electron pair. It has been produced by a photon passing close to the nucleus of an atom of the liquid and thus passing through the electric field associated with the charge on the nucleus. The lower triplet consists of a similar pair produced in the electric field of an atomic electron; in the process, the atomic electron is knocked on and forms the third track.

No. 4 Four-pronged interaction

(See Fig. 31)

Here you can see the conversion of kinetic energy into rest-mass energy, and rest-mass energy into electromagnetic energy.

A beam of negatively-charged pions enters the chamber from the lower side. These are particles produced in collisions involving protons from the accelerator. They have been separated out from the other particles coming from the target by the methods described in section 32.3. Most of the pions pass straight through the chamber and out the other end. One, however, comes within range of the strong interaction of the nucleus of a hydrogen atom, and four charged particles emerge from the resulting collision. These comprise a proton and pion *plus* two more pions, one negatively charged and the other positively charged. Some of the kinetic energy of the original pion has been transformed into rest-mass energy for the two additional particles. Measurements made on the curvatures of the tracks allow the momenta and hence energies of the particles to be determined. These measurements show that Einstein's famous equation $E = mc^2$ holds (Unit 4).

The process can be conveniently summarized in the form of an equation:

$$\pi^- + p \rightarrow \pi^- + p + \pi^+ + \pi^- \quad \dots\dots\dots\dots\dots\dots\dots\dots\dots (5)$$

In these equations, each particle is represented by a symbol: p for proton is already familiar to you, and the Greek letter π (pi) represents the pion. The + and − signs indicate the sign of the electric charge of the particle.

electron pair

energy transformations

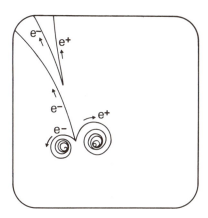

Figure 30   *Diagram for stereophotograph No. 3: electron pair and triplet.*

the production of pions

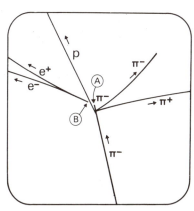

Figure 31   *Diagram for stereophotograph No. 4: four-pronged interaction.*

* *You first came across positrons in the Broadcast Notes of Unit 2.*

(It is always taken for granted that the proton has positive charge, so no one ever bothers to put a $^+$ sign on the p.) It is not necessary to specify *how much* negative or positive charge a particle has, because electric charge only comes in multiples of the charge on the electron. If the same symbol is used in denoting more than one particle, e.g. the symbol π in π^+ and π^-, then it is to be understood that the particles are more or less identical apart from the electric charge they carry.

You will notice that one of the π^-s from the reaction represented by equation 5 (partially obscured by the proton track) does not get very far before it interacts with another proton (at point A in Figure 31). It loses its charge to become neutral:

$$\pi^- + p \rightarrow \pi^0 + n \dots \dots \dots \dots \dots \dots \dots \quad (6)$$

(Note that a zero sign is needed to specify that the final pion is neutral. It is, however, customary to omit the zero sign for the neutron because it is such a well-known particle and can only be neutral.) Pions are unstable particles; they live only a very short time before decaying into something else. As you will learn in your main experiment for this Unit (Home Experiment No. 2), charged pions decay into muons; this happens in about 2×10^{-8} s. The π^0 lives for an even shorter time, 10^{-16} s. Before it can move any detectable distance (while it is essentially still at point A), it decays into two high-energy photons called gamma rays (represented by the symbol γ):

$$\pi^0 \rightarrow \gamma + \gamma \dots \dots \dots \dots \dots \dots \dots \dots \dots \quad (7)$$

the fast decay process
$\pi^0 \rightarrow \gamma + \gamma$

In this process, the rest-mass energy of the π^0 converts into energy in the form of two flashes of light, i.e. into electromagnetic energy.

What happens to one of the photons from the π^0 in your stereophotograph?

One of the photons produces an electron pair at point B in Figure 31 (the conversion of electromagnetic energy into rest-mass energy again).

No. 5 A neutron interaction

(See Fig. 32.)

A π^- enters the chamber and interacts with a proton to produce a π^0 and a neutron:

$$\pi^- + p \rightarrow \pi^0 + n \dots \dots \dots \dots \dots \dots \dots \quad (8)$$

This is the same as what happened at point A in Figure 31, but this time both γs from the decay of the π^0 leave the chamber without giving electron pairs. The neutron however collides with a proton and produces another particle—a negative pion. Electric charge is conserved by the neutron becoming a proton:

$$n + p \rightarrow p + p + \pi^- \dots \dots \dots \dots \dots \dots \quad (9)$$

At this point, we interrupt our commentary to make a general remark. As you will probably have gathered already from our study of the first five stereopictures, a bubble-chamber physicist can seemingly read an awful lot into a few trails of bubbles! How can he possibly be so confident in his interpretations? It is one thing for him to say that he knows the momentum of a track from its curvature, but how can he go on to assert that the track is that of a pion or proton?

The answer lies mostly in a detailed study of the interactions. Remember these interactions have to conserve both momentum and energy. Particles of the same momentum but different mass will have different energies.

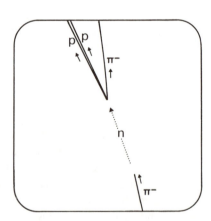

Figure 32 *Diagram for stereophotograph No. 5: a neutron interaction (an interaction is sometimes called a 'star' as it is on this reel).*

33

Very often, the requirement that the total energies before and after the collision (or decay) should be equal is sufficient to establish a unique choice of masses for the various particles. And even without such detailed information, a physicist can often spot helpful clues in the picture. We give one example: the number of bubbles produced per unit length along the path of the particle depends inversely upon the square of the particle's velocity; hence the lower the velocity the greater the number of bubbles.

dependence of bubble density on particle velocity

Take another look at stereophotograph No. 5. The π^- track from the neutron interaction hits on an electron which makes its own spiralling track. Study very closely the π^- and electron tracks. Which particle was the fastest, the pion or the electron? Which particle had the greater momentum, the pion or the electron?

The electron track has less bubbles per unit length and so corresponds to the faster particle. But it also has the smaller radius of curvature, and therefore the smaller momentum. Because momentum is defined as mass × velocity, it is obvious that for a particle to have a higher velocity, but a lower momentum, it must have a lower mass. This simple observation allows one to conclude immediately that the particle producing the pion track must have been heavier than that producing the electron track.

With experience, one learns to combine information from the number of bubbles with that from magnetic curvature to estimate the mass of each particle. These ionization estimates are very rough (about ± 30 per cent), but are often good enough to distinguish between various hypotheses regarding the identity of the particle, such as, for example, between electron and pion, or between pion and proton, since the masses are so different.

32.4.3   Conservation laws

Particles appearing and disappearing, some photons giving electron pairs, others not, some collisions producing pions, others not—what does it amount to? Before we move on to the last two stereopictures let us take stock.

Is the situation as chaotic as it might seem at first sight? No. The collision and decay processes are governed by strict rules of behaviour—*the conservation laws*. One of the tasks of the physicist is to seek and understand these laws.

conservation laws

We have already made explicit mention of three conservation laws in this Unit. What were they?

The law of conservation of electric charge is one such law—it is always strictly obeyed. The law of conservation of energy is another such law—it too is always obeyed (assuming of course, that the appropriate energy, given by the equation $E = mc^2$, is assigned to the rest-masses of all particles, and that a long time is available for measurement). Conservation of momentum is a third law that you know about (Unit 3). You probably did not realize it, but in our interpretation of the photographs we were also making use of yet another conservation law. If you look back over the various processes described in this section—and also the radioactive decays of Unit 31—you will note that the number of nucleons (i.e. the number of neutrons plus the number of protons) before and after the interaction is always the same. From this it might be concluded that there is a law of conservation of nucleons. This would not however be quite true (as you will see in the next stereopicture). You will remember from section 32.1 how we mentioned that many new elementary particles have

been discovered since the pion. Well, some of these new particles behave in certain respects like the nucleons. It becomes convenient to introduce a new name for referring to nucleons and to these new particles that behave like nucleons; they are collectively called—*the baryons*. Just as the neutron and proton are collectively called nucleons, so the nucleons and these new particles belong to the wider family of baryons. *The law of conservation of baryons says that although one baryon may change into another, the total number of baryons before and after the interaction must be the same.**

the general class of particles called baryons

the law of conservation of baryons

Pions are not baryons.

Look back over the interactions you have seen and decide whether a similar law of conservation of pions exists.

There is no equivalent law of conservation of pions. Particles that can be produced in any numbers in strong interactions, such as pions, belong to a class of particles called *mesons*.

the general class of particles called mesons

Now look at the next stereopicture.

No. 6 Associated production of strange particles

(See Fig. 33)

In this collision between an incoming pion and a proton you are introduced to *two* new particles, the Σ^- (pronounced 'sigma minus') and K$^+$ (pronounced 'kay plus'). The reaction can be written

$$\pi^- + p \rightarrow \Sigma^- + K^+ \quad \ldots\ldots\ldots\ldots\ldots\ldots (10)$$

The Σ^- particle, which is on the left of the picture, soon decays to a neutron and π^-, the π^- diving steeply away from you and out through the far wall of the chamber. Note that this is a decay *not* a collision; the Σ^- spontaneously breaks up into two decay fragments:

$$\Sigma^- \rightarrow \pi^- + n \quad \ldots\ldots\ldots\ldots\ldots\ldots (11)$$

Similarly the K$^+$ on the right decays:

$$K^+ \rightarrow \pi^+ + \pi^\circ \quad \ldots\ldots\ldots\ldots\ldots\ldots (12)$$

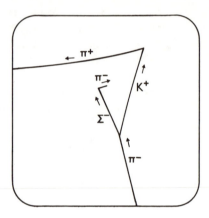

Figure 33 *Diagram for stereophotograph No. 6: associated production of strange particles.*

From these processes decide

(i) whether the Σ^- is a baryon or meson;

(ii) whether the K$^+$ is a baryon or meson.

In equation 10, the law of conservation of baryons requires that, corresponding to the single baryon present before the interaction (the proton), there should be one baryon after the interaction. Therefore *either* the Σ^- or the K$^+$ must be a baryon. Equation 11 tells us that the Σ^- is a baryon, since the neutron is, and the π^- is not. So the K$^+$ cannot be a baryon. This conclusion is confirmed by equation 12 where neither of the final particles is a baryon.

We now come to a very important observation. You may have been thinking that this photograph must be rather exceptional in that it has *two* new particles and not just one. Why did we not introduce you to these particles one at a time? Why not take, for example, the reaction $\pi^- + p \rightarrow \Sigma^- + \pi^+$? The reason is simple—*this reaction does not happen.*

---

* *More strictly speaking the law requires that the number of baryons minus the number of anti-baryons must remain constant. A discussion on anti-particles, though, is beyond the scope of this Course.*

Neither of the laws is violated. There must be some other reason for the absence of this reaction.

In collisions involving only nucleons or pions, it is found from the study of many interactions that the Σ^- is *only* produced in the company of a positively charged or neutral K meson as in equation 10. Indeed, there are many other reactions that appear on the face of it to be perfectly 'reasonable'—but they also never occur. Something seems to be preventing them from happening. But what can it be?

This type of situation suggests to a physicist that some kind of conservation law is at work. To see this, first take a look at the way a more familiar conservation law operates—the law of conservation of electric charge.

This law does *not* say that it is impossible to create extra units of electric charge. It says instead that if an amount of negative electric charge is to be created, then an equal amount of positive charge must be created at the same time.

Thus, for example, in equation 9, a negative electric charge for the pion could be created because at the same time the neutron acquired an equal positive charge and became a proton.

The impossibility of producing a single Σ-particle may be analogous to the impossibility of producing a single electric charge. Perhaps the Σ-particle has some hitherto unsuspected property which, like electric charge, must be conserved. The fact that Σ-particles are always produced in association with a K^+ (or a K°) meson would then be an indication that whatever this unsuspected property might be, the K^+ (or K°) meson has an equal and opposite amount of it. Thus the simultaneous production of a Σ and a K does not alter the net amount of this property.

It is now known that this hypothesis is correct; there *is* another property of elementary particles. It has no counterpart in the macroscopic world; it cannot be identified with mass, or with angular momentum or indeed with anything that can be easily visualized. But all the evidence is that elementary particles do possess this extra property. It has been given the rather colourful name—*strangeness*.

strangeness

The K^+ is assigned $+1$ unit of strangeness; the Σ^- has -1 unit. (As with electric charge, the sign convention is quite arbitrary. Also the size of the unit is decided only on the grounds of convenience. If thought desirable, one could have assigned, for example, -72 units of strangeness to the K^+ —as long as an equal and opposite amount of it was given to the Σ, i.e. $+72$ units. Incidentally, these units do not have any of the usual dimensions of Mass, Length and Time.) Pions, neutrons and protons have zero strangeness. *The law of conservation of strangeness states that strangeness is conserved in strong interactions.*

the law of conservation of strangeness

Thus in the reaction $\pi^- + p \rightarrow \Sigma^- + K^+$, the strangeness on the left-hand side is $0+0=0$ and on the right-hand side $-1+1=0$. In the non-existent reaction $\pi^- + p \rightarrow \Sigma^- + \pi^+$, the strangeness on the left-hand side would again be 0, but that on the right-hand side would be $-1+0=-1$. The reaction would violate the law of strangeness conservation, and so cannot occur.

But, you may object, what about the decay processes $\Sigma^- \rightarrow \pi^- + n$ and $K^+ \rightarrow \pi^+ + \pi^\circ$? These manifestly violate strangeness conservation. This is true, but let us draw your attention to the wording of the law; it specifically states 'in strong interactions'. It is found that the law does *not* apply to processes in which particles spontaneously decay without colliding with anything. Other laws apply to these processes, but not the strangeness conservation law.

a restriction on the validity of the law of conservation of strangeness

How will you know whether an interaction is strong or not? In what follows, any collision with a nucleon is caused by the strong interaction;

36

any decay process, on the other hand, is caused by the so-called weak interaction.*

Because the strangeness law is not obeyed in decay processes, it does not have the universal validity characteristic of, say, the law of conservation of electric charge. A little messy? Perhaps, but that's the way it is!

To see how this law operates, we shall consider a few more interactions. Take for example the following observed collision process involving a nucleus:

$$\pi^+ + p \rightarrow \Sigma^+ + K^+ \quad \ldots\ldots\ldots\ldots (13)$$

From what you already know, deduce the value of the strangeness of the Σ^+.

Because it is a collision process involving a nucleon, it is a strong interaction; the law of strangeness conservation can therefore be applied. The strangeness of the left-hand side is $0 + 0 = 0$. In order for the strangeness of the right-hand side to be zero also, the strangeness of the Σ^+ must be equal and opposite to that of the K^+. Like the Σ^-, therefore, the Σ^+ must have strangeness of -1.

For another example of the way the strangeness conservation law works, take a look at the final stereophotograph.

No. 7 A K^- meson interaction

(See Fig. 34.)

A K^- meson loses its kinetic energy through ionization and stops in the liquid. It is attracted to the nearest hydrogen nucleus because of the force exerted between their opposite electric charges. On coming within range of the strong interaction of this proton, the K^- meson undergoes the following reaction:

$$K^- + p \rightarrow \Sigma^+ + \pi^- \quad \ldots\ldots\ldots\ldots\ldots\ldots (14)$$

(Like the Σ^- in the previous stereophotograph, the Σ^+ subsequently decays to a nucleon and pion: $\Sigma^+ \rightarrow \pi^+ + n$.)

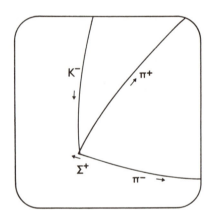

Figure 34   *Diagram for stereophotograph No. 7: a K^- meson interaction.*

From what you already know, deduce the strangeness of the K^- meson.

The K^- is involved in a strong interaction, so strangeness must be conserved. The strangeness of the right-hand side is $-1 + 0 = -1$. As the strangeness of the proton on the left-hand side is zero, that of the K^- must be -1.

Although the strangeness of the Σ^- and Σ^+ are the same, that of the K^- and K^+ are *opposite*. Thus according to this assignment, although the process $\pi^- + p \rightarrow \Sigma^- + K^+$ conserves strangeness, the very similar-looking process $\pi^- + p \rightarrow \Sigma^+ + K^-$ does *not*. (The left-hand side has zero strangeness, the right-hand side has $-1 + -1 = -2$.) Therefore, a consequence of the law of conservation of strangeness is that one can never hope to see a picture, analogous to stereophoto No. 6, in which the charges on the Σ and K are interchanged! This remarkable conclusion is found in practice to be fully vindicated. Although hundreds—perhaps thousands—of examples of $\Sigma^- K^+$ production have been studied, no one has ever come across an example of $\Sigma^+ K^-$ production.

an important distinction between the K^- and K^+

* *This is not a general rule, but it will be true of the collisions and decays we shall be considering subsequently in this Unit.*

37

Thus, having used a few reactions to establish the strangeness values of the new particles, one can then go on to use these values to make predictions concerning the occurrence or otherwise of all other possible reactions. We have given you two examples; there are many others we could cite. This predictive ability is one of the important features of the law of conservation of strangeness.

For further practice in handling baryon and strangeness assignments and in making predictions try SAQs 13 to 17.

32.5 The Theory of Elementary Particles

32.5.1 A classification scheme for elementary particles

Sifting through millions of interactions every year, physicists are continually discovering new particles and testing their behaviour. In ways similar to those you have just been using, they deduce for each new particle whether it is a baryon or meson, whether its strangeness is 0, -1, $+1$, -2, etc., as well as some other properties we have not been able to include in our brief introduction to the subject. For each new particle a catalogue of properties is compiled. Some of them have familiar everyday counterparts: mass, electric charge, etc.; others such as strangeness have no macroscopic counterpart, but are vital nonetheless to an understanding of the particle's behaviour. As accelerator energies increase, so it becomes possible to add more and more particles to the list.

But is this all there is to it—just a catalogue of particles and their properties? No, this is only the beginning. The next step is to classify them. Just as Mendeleev's Periodic Table brought order to the list of elements and their properties, so the physicist seeks to bring order to the list of elementary particles. This is no easy task—there are so many possible classification schemes one could think of. For example, should the particles be classified into baryons and mesons? If so one ought to separate n, p, Σ^-, from π^+, π^-, K$^+$, and K$^-$. Or should the classification be according to strangeness? In that case, n, p, and π^0 for example would go together (with strangeness zero), and so would Σ^-, Σ^+, and K$^-$ (with strangeness -1). If the classification is according to electric charge then p, Σ^+ and K$^+$ would be members of one grouping, and n, π^0, and K^0 members of another.

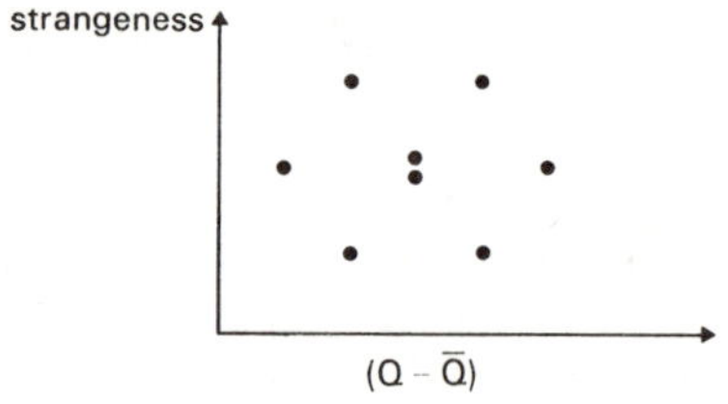

Figure 35 *According to the SU*3 *scheme, groupings of eight particles form a hexagonal array on a plot of strangeness versus* $(Q-\bar{Q})$.

In 1961, Gell-Mann in the United States and Ne'eman who was working at the time in England, simultaneously achieved a breakthrough with a theory called 'SU3'.*

the SU3 theory

According to this scheme the particles divide themselves into groupings of one, eight and ten particles each, where the particles in each grouping are characterized by having certain properties in common. One of these properties, for example, is spin. Particles may spin much in the same way as the Earth spins about an axis passing through its centre. In order for particles to belong to the same grouping, they must each spin with the same angular momentum. Likewise particles of the same grouping must be either all baryons or all mesons.

the particles in an SU3 grouping have certain properties in common—but in other ways they differ from each other

However, in some respects the particles in a grouping are *not* identical. Indeed, the way in which the particles *differ* from each other is especially interesting. This can best be seen by displaying the members of a grouping on a two-dimensional grid or graph, on which is plotted two properties that the particles do not have in common. When each particle is assigned to its location on the graph, the grouping forms a characteristic pattern. All groupings of eight particles have the pattern seen in Figure 35. The ordinate gives the value of the strangeness of the particle, and plotted along the abscissa is the value of another property of elementary particles: $(Q-\bar{Q})$. Though we hesitate to introduce yet another property, we feel

another property of particles: $(Q-\bar{Q})$

---

* SU3 *is the name of a mathematical 'group'. Unfortunately, without a knowledge of group theory one cannot really understand the significance of the name—'Special Unitary group 3'.*

justified in this case because its meaning is very easy to understand. This meaning is best illustrated by an example.

There are two types of nucleon—the proton and the neutron. One is positively charged, the other neutral. The 'average charge', $\overline{Q}$, of the nucleons is then given by

$$\overline{Q} = (1+0)/2 = +\tfrac{1}{2}$$

Q is simply the charge on the individual particle. Therefore in the case of the proton, for which $Q = +1$,

$$(Q-\overline{Q}) = (1-\tfrac{1}{2}) = +\tfrac{1}{2}$$

What is the value of $(Q-\overline{Q})$ for a neutron?

For a neutron $Q=0$, so

$$(Q-\overline{Q}) = (0-\tfrac{1}{2}) = -\tfrac{1}{2}$$

Given that there are three types of Σ-particles, Σ^+, Σ^0, and Σ^-, what is the value of $\overline{Q}$ for Σ-particles?
What is the value of $(Q-\overline{Q})$ for each individual Σ-particle?

The three types of Σ-particle have charges $+1$, 0 and -1 respectively. Therefore

$\overline{Q} = (+1 +0 -1)/3 = 0/3 = 0$.
Thus $(Q-\overline{Q})$ has the following values—
For the Σ^+: $(+1 -0) = +1$
For the Σ^0: $(0-0) = 0$
For the Σ^-: $(-1 -0) = -1$

For further practice in assigning values of $(Q-\overline{Q})$ to particles, try SAQ 18.

The idea is that, although the particles in each grouping have *some* properties in common (e.g. they may all be baryons), the values of strangeness and $(Q-\overline{Q})$ differ. When the particles are displayed on a plot of strangeness versus $(Q-\overline{Q})$, the groupings of eight form a hexagonal array, with two particles at the centre (see Fig. 35). The particular grouping to which the nucleons belong is shown in Figure 36. You can see that they are lumped together with the Σ-particles—and also the Λ and Ξ particles, which you met in SAQ 13.

Using the values of strangeness and $(Q-\overline{Q})$ which you have deduced for these particles in SAQs 13 and 18, check that they are in their correct locations on this graph.

Figure 36 shows that under this scheme, nucleons, despite their abundance in the world, have no privileged position; *the Λ, Σ, and Ξ particles are every bit as important, basic and essential as the nucleons.*

Figure 37 shows the general form of the array corresponding to a grouping of ten particles; it is triangular. An example of a grouping of particles that almost fits this form of array is given in Figure 38.

Without worrying about the symbols denoting the names of the particles, do you notice what is different between Figures 37 and 38?

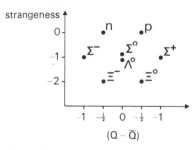

Figure 36   A particular grouping of eight particles.

The odd thing about Figure 38 is that the apex of the triangle is missing—a tenth particle is needed to complete the array. This is how it was in 1962 regarding this grouping.

Can you see a resemblance between this situation and that in which Mendeleev found himself when compiling the Periodic Table of elements? He too discerned groupings; he too found that certain members were missing. Mendeleev left gaps for undiscovered elements. Indeed he went further and *predicted* the existence of these elements and, from the location of the gaps, specified the likely properties of these elements.

40

Can we do the same? Can we predict, on the basis of Figure 38, the existence of a new particle, and from the location of the gap, specify in advance what its properties might be? Or is the gap merely evidence that the SU3 theory is wrong?

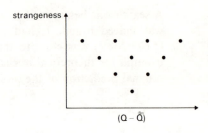

Take another look at Figure 38 and predict the following properties of a possible tenth particle:

  (i) the value of its strangeness;

  (ii) the value of $(Q - \overline{Q})$;

  (iii) the value(s) of its electric charge.
  (To be able to hazard a guess as to what the last property might be, you need to study the values of the electric charges on the other particles. Note that the two plus signs associated with the extreme right-hand member of the top row mean that it has a positive charge of two units—in this respect it is unusual.)

Figure 37 According to the SU3 scheme groupings of ten particles form a triangular array on a plot of strangeness versus $(Q - \overline{Q})$.

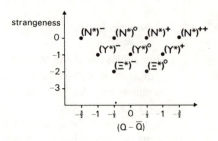

Figure 38 Can these particles be members of a grouping of ten particles?

From the position of the gap in Figure 38, the missing particle clearly must have a strangeness of -3 units, and a value for $(Q - \overline{Q})$ of zero. Its electric charge is -1 unit; this can be inferred from the fact that the left-hand member of each row is negatively charged (and the missing particle is the left-hand member of a row of one!) In this way, the suggested particle came to be called the *omega-minus* particle (Ω^-).

omega-minus

How about the mass of the Ω^-? Do the masses of the other particles in the array hold out any clues as to what this might be? Well, it is found that all members of a horizontal row have the same mass to within a few MeV (of rest-mass energy). The masses of the particles in each successive row are as follows:

N*	1 238	MeV
Y*	1 385	MeV
$\varXi$*	1 530	MeV
Ω^-	?	

Can you make a reasonable sort of guess as to what the mass of the Ω^- might be? (Because of the differences between individual members of a horizontal row, you cannot expect to get closer than a few MeV. Note also we are *only* asking for a guess.)

The difference between the masses characteristic of the first two rows is $1\,385 - 1\,238 = 147$ MeV; the difference between the second and the third is $1\,530 - 1\,385 = 145$ MeV. These two differences are the same (to within the few MeV accuracy to which we are working). It is therefore tempting to suggest that the difference in mass between the last two rows (i.e. between the $\varXi$* and the Ω^-) might be the same—about 145 MeV. That would make the mass of the Ω^-, $1\,530 + 145 = 1\,675$ MeV (to within a few MeV).

To summarize: in order to complete the array, a new particle is needed with a mass of about 1 675 MeV, carrying negative charge only, and possessing an unprecedented -3 units of strangeness. Does the particle exist? This question was to be the acid test of the SU3 theory.

A search was begun in 1962 and after two years the elusive omega-minus was indeed found. It had all the properties predicted—even its mass (1 672 MeV) was close to that expected. The discovery was hailed as a triumph for theoretical nuclear physics—one matched only by Yukawa's original prediction of the pion.*

a demonstration of the predictive power of the SU3 theory

At this point you might like to try SAQ 19.

32.5.2 A forward look

But a classification scheme, no matter how neat or how useful as an indicator of undiscovered particles, can hardly be regarded as an ultimate explanation. As you already know, the Periodic Table of elements, together with the atomic energy levels revealed by spectroscopy, were not in themselves explanations of anything—they were merely the outward signs of something much deeper and more fundamental. An explanation of the elements came only with an understanding of atomic structure in terms of nuclei and electrons (Units 6, 7, and 30). Likewise, the existence of different kinds of nuclei, together with their energy levels as revealed by nuclear spectroscopy and mass measurements, could only be explained in terms of their internal structure (Unit 31).

a similarity with nuclear spectroscopy

Now, for a third time, the physicist is faced with a similar situation. He knows the classification scheme of the elementary particles. He also knows some of their masses, and these can be regarded as 'energy levels'. The main difference between elementary-particle energy levels and those of atoms and of nuclei appears to be one of scale. Whereas differences in atomic energy levels are reckoned in terms of a few eV (Unit 6), and those between nuclear levels are of the order of a few MeV (Unit 31), the elementary particles differ in rest-mass energy by tens or hundreds of MeV.

So where do we go from here? Are the elementary particles *really* elementary or is their classification scheme and energy-level structure an outward sign of something deeper? Could the elementary particles be merely composite structures of something yet more basic? Is the proton's 'indivisibility' destined to suffer the same fate as that of the atom and the nucleus?

Does SU3 point the way to a new break-through in our understanding of the ultimate structure of matter?

The SU3 classification scheme holds out a tantalizing promise. The scheme is essentially based on three basic 'states' of strongly interacting particles. (The fact that there are three has to do with the number 3 in the name SU3.) These 'states' have been called *quarks*. But what exactly do we mean by the word 'states' as used in this context? It is very difficult to say. It is perfectly possible that they refer to nothing more than a mathematical concept. But many physicists hope that there is more to them than that; they hope that quarks will turn out to be real physical particles. (It is a little like the situation you encountered with waves. The concept of waves can be applied both to physical waves requiring a medium, like sound or earthquake waves, and also to waves that are purely mathematical like the probability waves of Unit 29 which require no medium and in that sense are not physical.)

quarks

* The story of the omega-minus is told in this Unit's radio programme by those most closely connected with it—Dr. M. Gell-Mann and Dr. Y. Ne'eman, and also Dr. N. P. Samios who was one of the team of 33 physicists who made the discovery.

The idea that quarks might be physically-detectable particles is an attractive one. The existence of quarks would allow all two hundred so-called 'elementary' particles to be described as composite structures of just three basic building blocks—the three types of quark. According to this scheme a proton would consist of three quarks tightly bound together.

Unfortunately, it has so far proved impossible to identify these quarks with any of the established particles—for one thing, quarks are expected to have electric charges of $\frac{1}{3}e$ or $\frac{2}{3}e$ (where e is the magnitude of the charge on the electron) and all known particles have charges that are integral multiples of e. A possible reason why quarks have so far eluded detection could be that they are exceedingly massive; the energies of present-day accelerators might be insufficient to produce them. But how, you may ask, can they be so massive when three of them add up to only one proton? Odd as it may seem, there is no contradiction here. Three quarks, each much more massive than a proton, could bind together with such colossally strong forces that the binding energy, and hence mass defect (Unit 31), would wipe out almost the whole of the mass of the three separate quarks! What we know as the proton mass may simply represent a tiny remnant left over after the binding energy has been subtracted from the masses of the quarks. Such a colossally strong force would also account for why no one has yet managed to break a proton apart into its separate quark constituents.

This also raises the question of whether the nuclear force might also be merely a remnant—a remnant of a much stronger quark-quark force acting in the interior of the nucleon. It is an interesting thought. After all, we know that the electrostatic attraction inside an atom, between its nucleus and its electrons, can give rise to an external force that binds atoms together to form molecules and solids and liquids. This inter-atomic force is complicated in that it has a short range and becomes repulsive at very small distances. This is very like the force acting between nucleons—so much so that you used liquid drops in Unit 31 to simulate the behaviour of nuclei. The complicated inter-atomic force has its origin in the simpler electrostatic force operating within the atom; perhaps the complicated nuclear force will one day be explained in terms of a simpler quark-quark force operating within the nucleon.

One final point we ought to mention. You saw in the previous Unit how, in order to understand the nuclear fusion processes occurring in the Sun (processes upon which our life on Earth depends), one had to study the behaviour of individual nuclei in a laboratory. In this Unit, you have been concerned with the physics of the very small; paradoxically it too may be connected with the physics of the very large. When a large star exhausts its nuclear fuel, it contracts under the influence of its own gravitational force. In the process, the kinetic energies of the particles in the star are believed to increase to values of the order of GeV. At this stage in stellar development, mesons and other elementary particles might be produced and high-energy nuclear physics would then take over on the grand scale. So particle accelerators and detectors, such as the ones at CERN, may be regarded, if you like, as man's ingenious means of looking into the fiery interior of a star in its final death throes.

In this Unit, we have covered much ground. We began by considering the nuclear force and this took us into the realm of elementary particles. We pointed out that the familiar neutron and proton were no more basic than many other particles, and indeed one is now no longer sure that these or any other known particles are really fundamental at all. And as for the nuclear force, it too may not be as basic as we thought at first. With the bigger accelerators, at present under construction, will quarks be discovered? We do not know. Will the Open University in future years be sending your successors stereophotographs of protons being smashed up into their constituents? We shall have to wait and see.

Summary

A new model for describing forces is introduced. According to this model, a force acting between two or more objects may be represented by the exchange of some intermediary between them.

This model is applied to the nuclear force. From the uncertainty relation and the known range of the force, the mass of the intermediary, called a pion, can be determined; it is 273 times the mass of an electron.

When sufficient energy is available, as it is in a very violent nuclear collision, some of it may transform into the rest-mass energy of the intermediary pion, which can thereby take on a real and separate existence.

Other particles have also been discovered. The exchange model of the force must therefore be developed further to allow the nucleon to exchange additional heavier particles during close approaches.

The study of the nuclear force, therefore, becomes the study of elementary particles.

To produce these particles, protons have to be accelerated to high energies. This is done in three stages. First of all the protons are subjected to an intense electric field. Secondly, they pass through the drift tubes of a linear accelerator and are accelerated by an alternating electric field. Finally they are accelerated to the highest energy in a synchrotron. In this machine, they are steered on a circular path by a magnetic field, and accelerated by electric fields.

The protons emerge in batches and are made to strike a target. The particles produced in these collisions are separated according to mass and momentum by a combination of slits and magnetic and electric fields.

The behaviour of the particles can be observed with a bubble chamber. In this instrument, pressure is reduced on a liquid so putting the liquid into a superheated condition. Bubbles grow on 'hot spots' left by the passage of charged particles. These can be photographed.

Through a study of stereopictures, some insight is given into the analysis of bubble-chamber tracks. The analysis of such photographs reveals that the interactions take place according to certain conservation laws— conservation laws of energy, momentum, electric charge, etc. It becomes necessary to introduce entirely new properties for the particles—properties that have no macroscopic counterpart. One such property, strangeness, is found to be conserved in strong interactions.

It has been found useful to classify the elementary particles using the 'SU3' theory. According to this scheme, the particles are displayed in arrays. The appearance of gaps in these arrays has led to the prediction of new particles—the most significant being that of the omega-minus particle.

The Unit ends on a note of speculation. Does the SU3 theory point the way to a much deeper understanding of the structure of matter? Are protons and other so-called elementary particles built up from a yet more fundamental entity—the quark?

Book List

For those who wish to pursue the subject of elementary particles a little further we suggest you consult the following books:

K. W. Ford, *The World of Elementary Particles*, 1963, Blaisdell.

H. S. W. Massey, *The New Age in Physics*, 1966, Elek.

Self-Assessment Questions

Section 1.3

Question 1 (*Objective 1*)

(a) *True*   (b) *False*

The nucleus of any element is an elementary particle.

Section 2.3

Question 2 (*Objective 2*)

(a) *True*   (b) *False*

In a linear accelerator the voltage on the drift tubes increases progressively from one end of the accelerator to the other.

Question 3 (*Objective 2*)

(a) *True*   (b) *False*

Hydrogen gas is led into one end of a linear accelerator and as the hydrogen atoms are accelerated towards the first drift tube they are stripped of their electrons.

Section 2.4

Question 4 (*Objective 2*)

(a) *True*   (b) *False*

In a synchrotron the purpose of the magnetic field is to accelerate the protons in a direction tangential to the synchrotron ring.

Question 5 (*Objective 2*)

(a) *True*   (b) *False*

Two synchrotrons, possessing magnetic fields of the same maximum strength, are such that one machine has a circumference twice that of the other. The maximum momentum of the protons from this machine is twice that of the other.

Question 6 (*Objective 2*)

(a) *True*   (b) *False*

In a synchrotron, the protons are repeatedly reflected at the walls of the circular vacuum chamber, and this helps to constrain them to move on a circular path.

Question 7 (*Objective 2*)

How much energy, expressed in joules, is contained in each batch of 10^{12} protons accelerated in the CERN machine to 28 GeV?

($1eV = 1.6 \times 10^{-19}$ J)

Section 3.1

Question 8 (*Objective 3*)

A particle travels from a point A to a point B at constant speed through a magnetic field. It is required to send a second particle in the reverse direction from B to A over the identical path. The two particles each have an electric charge of magnitude one unit.

The particles are such that:

(i)   their unit charges have (a) the same sign, (b) opposite sign?

(ii) their momenta are (a) the same, (b) not necessarily the same?

(iii) their masses are (a) the same, (b) not necessarily the same?

(iv) their velocities are (a) the same, (b) not necessarily the same?

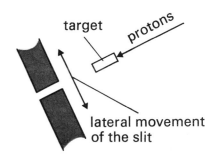

Figure 39 Diagram illustrating the direction of movement of the first slit considered in SAQ 9.

Question 9 (*Objective 3*)

Turn to Figure 23. When the slit closest to the target is displaced laterally, i.e. in the direction indicated in Figure 39, there is a gradual change in the number of particles passing through the slit. Likewise, when the second slit is displaced laterally there is a gradual change in the number of particles this slit allows through. However, a lateral displacement of the final slit gives rise to an irregular variation as in Figure 40. How do you explain these observations?

Section 3.2

Question 10 (*Objective 2*)

The pressure on the liquid in a bubble chamber is released (a) before, (b) during, or (c) after the passage of the particles?

Question 11 (*Objective 2*)

In a bubble chamber, there should be no irregularities in the walls of the container, otherwise boiling will occur on these irregularities and not along the paths of the charged particles.

Question 12 (*Objective 2*)

The lights used to illuminate the tracks in a bubble chamber are flashed at the instant the particles pass through the chamber.

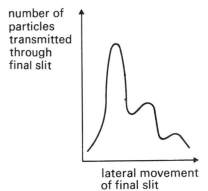

Figure 40 The variation in the number of particles transmitted by the third slit (in Figure 23) when this slit is moved laterally.

(a)   (b)
True   *False*

(a)   (b)
True   *False*

Self-Assessment Questions

Section 4.3

Question 13 (*Objective 5*)

In the text, you have been introduced to several particles. Some have been shown to be baryons, others mesons. Moreover, you have seen how their strangeness values are deduced. This information is summarized in Table 1.

Table 1

strangeness	baryon	meson
+2		
+1		K⁺K°
0	p n	$\pi^+\pi^-\pi^\circ$
−1	$\Sigma^-\Sigma^+$	K⁻
−2		

We shall quote six examples of strong interactions which have been observed, each involving a particle new to you. For each of the six new particles (shown in brackets alongside the reaction), deduce from the appropriate conservation laws whether it is a baryon or meson and what the value of its strangeness is. Record your answer by entering the symbol for the particle in the appropriate place in the table. (It is probably best to do this in pencil at first because you need a correct (!) version of Table 1 in order to be able to answer the next question.) Note that in the fourth reaction, the particle $\overline{K^0}$ (pronounced 'kay nought bar') is not to be confused with another particle, the K^0, to which you have already been introduced.

$$\pi^- + p \rightarrow n + \eta^0 \qquad (\eta^0)$$
$$K^- + p \rightarrow \Sigma^0 + \pi^0 \qquad (\Sigma^0)$$
$$K^- + p \rightarrow \Xi^0 + K^0 + \pi^0 \qquad (\Xi^0)$$
$$K^- + p \rightarrow \overline{K^0} + n + \pi^0 \qquad (\overline{K^0})$$
$$p + n \rightarrow \Lambda^0 + K^+ + n \qquad (\Lambda^0)$$
$$K^- + p \rightarrow \Xi^- + K^+ \qquad (\Xi^-)$$

Question 14 (*Objective 5*)

In order to attempt this question, you need to be able to refer to a correctly filled-in version of Table 1. So before reading on, check your answers to Question 13, and amend Table 1 as necessary.

We shall propose some hypothetical reactions and you are to decide for each whether the reaction proceeds as a strong interaction or not. If *not*, indicate which law(s) is(are) violated.

	YES	*because it violates the law of conservation of*		
		strangeness	*baryons*	*electric charge*
$\varXi^- + p \rightarrow \varLambda^0 + \varLambda^0$				
$K^- + p \rightarrow \varLambda^0 + \overline{K^0}$				
$n + p \rightarrow \varSigma^+ + K^0$				
$\pi^- + p \rightarrow \varSigma^0 + K^0 + \pi^+$				
$p + p \rightarrow \eta^0 + \pi^+ + n + \pi^0$				
$\varSigma^- + n \rightarrow \varLambda^0 + \pi^-$				
$\varLambda^0 \rightarrow p + \pi^-$				
$\varXi^0 + n \rightarrow \overline{K^0} + \varSigma^+ + \pi^-$				
$K^+ + p \rightarrow K^0 + \overline{K^0} + K^+ + \pi^+ + n$				
$K^+ + n \rightarrow \varXi^0 + \overline{K^0} + \pi^+$				
$\pi^- + p \rightarrow \varXi^- + \overline{K^0} + \overline{K^0} + n + \pi^+$				
$\pi^+ + n \rightarrow \varSigma^+ + K^0$				
$p + p \rightarrow \varSigma^+ + K^+ + n$				

Is it a strong interaction?

Question 15 (*Objectives 4 and 5*)

(In answering this question, refer as necessary to Table 1 of Question 13.)

A K⁻ meson interacts with a proton at point A in Figure 41 and produces a π^+, a π^-, and one neutral particle. This neutral particle decays at point B into a positive and a negative particle.

(i)   What is the strangeness of the neutral particle?

(ii) Is the neutral particle a baryon or meson?

(iii) Is the appearance of the two tracks coming away from point B consistent with the interpretation that the neutral particle decays into two particles of equal mass?

Question 16 (*Objectives 4 and 5*)

(In answering this question, refer as necessary to Table 1 of Question 13.)

A K⁻ meson interacts with a proton at point A in Figure 42. Coming from the interaction is a proton, a π^-, and one neutral particle. The neutral particle decays at point B into a π^+ and a negative particle, which for the time being we call 'X⁻'.

(i)   What is the strangeness of the neutral particle?

(ii) Is the negative particle, X⁻, a baryon or meson?

(iii) Can you deduce the strangeness of the negative particle, X⁻, by applying the law of conservation of strangeness first at A and then at B?

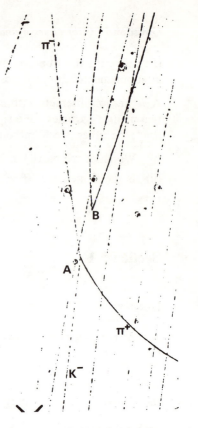

Figure 41   *A bubble-chamber photograph used in connection with SAQ 15.*

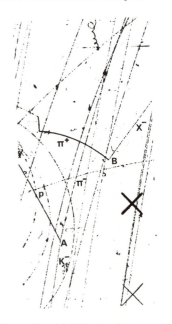

Figure 42   *A bubble-chamber photograph used in connection with SAQ 16.*

Question 17 (*Objectives 4 and 5*)

(In answering this question, refer as necessary to Table 1 of Question 13—as supplemented by your answer to Question 13.)

A π^- meson interacts with a proton at point A in Figure 43 to produce two neutral particles. One is a Λ^0 and this decays at B into a proton and a π^-; the other is a K meson which decays at C into a π^+ and a π^-.

(i) Which of the particles, X or Y, is the proton from the decay of Λ^0?

(ii) Is the K meson a K^0 or a $\overline{K^0}$?

Section 5.1

Question 18 (*Objective 6*)

Just as the 'nucleons' consist of (n, p), the 'sigmas' of (Σ^+, Σ^0, Σ^-), and the 'pions' of (π^+, π^0, π^-), so there are other particles e.g. (Ξ^-, Ξ^0); (K^+, K^0); ($\overline{K^0}$, K^-), (η^0) and (Λ^0). Note that there are two types of K with opposite strangeness so these have to be separated. Note also that the η^0 and Λ^0 have no charged counterparts.

Calculate the value of ($Q-\overline{Q}$) for each individual particle (where Q and $\overline{Q}$ are defined in the text).

	$(Q-\overline{Q})$		$(Q-\overline{Q})$
Ξ^-		$\overline{K^0}$	
Ξ^0		K^-	
K^+		η^0	
K^0		Λ^0	

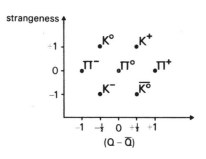

Figure 43   *A bubble-chamber photograph used in connection with SAQ 17.*

Question 19 (*Objective 7*)

The seven particles in Figure 44 almost make up an SU3 grouping of eight. What are the values of the strangeness and ($Q-\overline{Q}$) of the missing member?

Figure 44   *A grouping of particles that almost makes up an array of eight. (SAQ 19.)*

Question 1

Answer (b)

Comment

An elementary particle is one that cannot be broken down into more basic component parts. The nuclei of elements (except hydrogen) can be split up into their constituent neutrons and protons.

Question 2

Answer (b)

Comment

The drift tubes in Figure 9 are connected to the same alternating voltage supply, so their voltages must be the same.

Question 3

Answer (b)

Comment

The hydrogen nuclei are stripped of their electrons in the proton source, not in the linear accelerator. Until the electrons are removed, there is no way of accelerating the hydrogen atoms. The charge on the ionized atom is the 'handle' by which the electric field can get hold of the ion and accelerate it.

Question 4

Answer (b)

Comment

The acceleration is certainly in a direction tangential to the synchrotron ring, but it is produced by an electric not a magnetic field.

Question 5

Answer (a)

Comment

The circumference of a ring is $2\pi R$, where R is the radius of the ring. If one machine has twice the circumference of the other, it must have twice the radius. The path of the particles must therefore have twice the radius of curvature, and hence they must have twice the momentum. (Remember, for the same magnetic field, momentum is proportional to the radius of curvature.)

Question 6

Answer (b)

Comment

The protons are constrained to move on a circular path by the magnetic field, not by the vacuum chamber. The latter is there in order to allow the protons to move in a vacuum and so to minimize the number of protons that are scattered out of the machine before full energy has been reached.

Question 7

Answer 4 500 J

Comment

The final energy of each proton is 28 GeV i.e. 28×10^9 eV. This is the same as $(28 \times 10^9) \times (1.6 \times 10^{-19})$ joules. A batch of 10^{12} protons has energy

$$(28 \times 10^9) \times (1.6 \times 10^{-19}) \times 10^{12} \text{ J} = 4\ 500 \text{ J}$$

This is about the same as the kinetic energy of a brick dropped from the top of the Post Office Tower in London. The energy is sufficiently great that one would not want passers-by to be at the foot of the tower when the brick lands, but then again the energy is not particularly spectacular. What *is* remarkable about the energy delivered by the synchrotron is that it is concentrated into a tiny speck of matter, with a mass equivalent to 10^{-15} of a brick!

Question 8

Answer (i) b; (ii) a; (iii) b; (iv) b

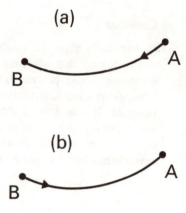

Comment

As far as electric charge is concerned, a charge transferred from A to B is equivalent to an equal but opposite charge transferred from B to A. Thus, if the particle moving from A to B in Figure 45 (a) is deflected to the right, an oppositely charged particle is needed for the journey from B to A, if it is to be deflected to the left (see Fig. 45 (b)). In order that the radius of curvature should be the same, the particle travelling from B to A must have the same momentum as that going from A to B. As long as the momentum (*mv*) is the same, it does not matter whether the mass or the velocity are the same.

Figure 45   See the answer to SAQ 8.

Question 9

Answer

Particles are emitted from the target in all directions, so a lateral displacement of the first slit should give a smooth variation in the number of particles passing through it. Likewise, the particles passing through the field of the bending magnet have a continuous range of momentum values; thus a lateral displacement of the second slit also gives a smooth variation. However, the final selection of particles by the electric field is according to mass. Although there might be several kinds of particle in the final beam, their masses are restricted to certain values e.g. those of the pion, proton, K meson, etc. Thus, a lateral displacement of the final slit reveals a series of peaks, each one corresponding to one of these types of particle.

Question 10

Answer (a)

Comment

The pressure on the liquid is released before the arrival of the particles so that the liquid is already in a superheated state. This is because the energy deposited by the particle rapidly disperses away from the path of the particle; unless the bubbles start to grow immediately while the energy is still concentrated along the path, they cannot grow at all.

In the TV programme, you will be shown a display on a cathode-ray oscilloscope which will make clearer the relationship in time of the arrival of the beam, the flashing of the lights, and the variation of the pressure on the liquid.

Question 11

Answer (b)

Comment

In the early days of bubble-chamber development (the early 1950s) it was feared that spurious boiling on irregularities in the walls of the container would prevent boiling along the paths of charged particles. Scrupulous care was therefore taken to keep all surfaces smooth and rounded. It was soon discovered, however, that such precautions were not necessary. Although in modern chambers the walls are kept smooth so as to prevent unnecessary spurious boiling, this type of boiling does nevertheless occur to some extent—as you saw in stereophotograph No. 1.

Question 12

Answer (b)

Comment

There must be a short delay (a millisecond or so) between the passage of the particles and the taking of the photograph, in order that the bubbles may have time to grow to a visible size.

Question 13

Answer

After making your additions, Table 1 should look like this:

strangeness	baryon	meson
+2		
+1		K^+K^0
0	p n	$\pi^+\pi^-\pi^0\eta^0$
−1	$\Sigma^-\Sigma^+\Sigma^0\Lambda^0$	$K^-\overline{K^0}$
−2	$\Xi^0\Xi^-$	

Comment

While getting used to the procedure of checking a reaction for strangeness or baryon conservation, it is a good idea to draw up a table under each reaction with a row in which the strangeness value of each particle is written, and another for indicating which particles are baryons (by putting '1' for a baryon and 'O' for a meson). It is then an easy matter to balance both sides of each equation. We show how this is done for the first and sixth reactions:

$$\pi^-+p \rightarrow n+\eta^0$$

strangeness	$0+0 = 0+s$	$\therefore s=0$
baryon	$0+1 = 1+b$	$\therefore b=0$

$$K^-+p \rightarrow \Xi^-+K^+$$

strangeness	$-1+0 = s+1$	$\therefore s=-2$
baryon	$0+1 = b+0$	$\therefore b=1$

(In Question 14 you will also be required to check that the reaction conserves electric charge. You might therefore wish to include a third row under the reaction for writing down the electric charge of each particle.)

Question 14

Answer

Your completed table should look like this:

Is it a strong interaction?

	YES	NO because it violates the law of conservation of:		
		strangeness	*baryons*	*electric charge*
$\Xi^- + p \rightarrow \Lambda^0 + \Lambda^0$	X			
$K^- + p \rightarrow \Lambda^0 + \overline{K^0}$		X		
$n + p \rightarrow \Sigma^+ + K^0$			X	
$\pi^- + p \rightarrow \Sigma^0 + K^0 + \pi^+$				X
$p + p \rightarrow \eta^0 + \pi^+ + n + \pi^0$			X	X
$\Sigma^- + n \rightarrow \Lambda^0 + \pi^-$			X	
$\Lambda^0 \rightarrow p + \pi^-$		X		
$\Xi^0 + n \rightarrow \overline{K^0} + \Sigma^+ + \pi^-$			X	
$K^+ + p \rightarrow K^0 + \overline{K^0} + K^+ + \pi^+ + n$	X			
$K^+ + n \rightarrow \Xi^0 + \overline{K^0} + \pi^+$		X		
$\pi^- + p \rightarrow \Xi^- + \overline{K^0} + \overline{K^0} + n + \pi^+$		X	X	
$\pi^+ + n \rightarrow \Sigma^+ + K^0$	X			
$p + p \rightarrow \Sigma^+ + K^+ + n$	X			

Comment

If you did not fare very well, we suggest you come back to this question in a day or two and have a second go at it, covering up your previous answers. There is really nothing very difficult about it—as long as you are not careless!

Question 15

Answer

(i)   The strangeness is -1.
(ii) The particle is a baryon.
(iii) The two particles do *not* have equal mass.

Comment

The reaction at point A can be written:

$$K^- + p \rightarrow \pi^+ + \pi^- + X^0$$

where the unknown neutral particle is being denoted by X^0. Conservation of strangeness requires the X^0 to have the same strangeness as the K^- i.e. -1. Conservation of baryons requires the X^0 to be a baryon.

The two particles from the decay of the X^0 are such that the positively-charged particle on the right has the greater momentum (judged from its greater radius of curvature in the magnetic field), but the lower velocity (judged from the fact that it has more bubbles per unit length than the negatively-charged particle on the left). It can only have a greater momentum (mv) and a lower velocity (v), if its mass (m) is greater than that of the other particle. In fact, the neutral particle is a Λ^0, and it decays as follows:

$$\Lambda^0 \rightarrow p + \pi^-$$

Question 16

Answer

(i)   The strangeness is -1.
(ii) Particle X^- is a meson.
(iii) No.

Comment

The reaction at point A can be written:

$$K^- + p \rightarrow p + \pi^- + Z^0$$

where the unknown neutral particle is being denoted by Z^0. Conservation of strangeness requires the Z^0 to have the same strangeness as the K^- i.e. -1. Conservation of baryons at point A requires the Z^0 to be a meson. At point B the Z^0 decays thus:

$$Z^0 \rightarrow \pi^+ + X^-$$

As the Z^0 and π^+ are both mesons, conservation of baryons requires X^- to be a meson too. Although the strangeness of the π^+ is known to be zero and that of the Z^0 has been found to be -1, it is *not* possible to conclude from strangeness conservation what the strangeness value of the X^- might be. This is because the decay of the Z^0 is caused by the *weak* interaction, not by the strong interaction, and so the law of conservation of strangeness does not apply to the reaction at point B (though of course the laws of conservation of baryons and electric charge *do* apply, as always).

In point of fact the 'Z^0' is a neutral K meson and the 'X^-' is a π^-.

Question 17

Answer

(i) Particle X.
(ii) K^0.

Comment

The various electrons in the picture, as well as the incoming particles, which you have been told are negatively-charged pions, are seen to curve to the right. The proton from point B must be positively charged and so is the one that curves in the opposite direction, i.e. to the left. (If you have difficulty seeing the curvature—which is admittedly small—you should place the edge of a straight ruler alongside the track.) The reaction at point A is either:

$$\pi^- + p \rightarrow \Lambda^0 + K^0$$
$$\text{or } \pi^- + p \rightarrow \Lambda^0 + \overline{K^0}$$

By looking up the strangeness values of these particles in your supplemented version of Table 1 of Question 13, you should easily be able to verify that only the first of these equations satisfies strangeness conservation. Therefore the particle must be a K^0, and not a $\overline{K^0}$.

Question 18

Answer

	$(Q - \overline{Q})$		$(Q - \overline{Q})$
Ξ^-	$-\frac{1}{2}$	$\overline{K^0}$	$+\frac{1}{2}$
Ξ^0	$+\frac{1}{2}$	K^-	$-\frac{1}{2}$
K^+	$+\frac{1}{2}$	η^0	0
K^0	$-\frac{1}{2}$	Λ^0	0

Comment

The charge values, Q, for the Ξ^- and Ξ^0 are -1 and 0 respectively. The 'average charge', $\overline{Q}$, for these two particles is $(-1+0)/2 = -\frac{1}{2}$

$$\therefore \text{For the } \Xi^-, (Q - \overline{Q}) = (-1) - (-\tfrac{1}{2}) = -1 + \tfrac{1}{2} = -\tfrac{1}{2}$$
$$\text{and for the } \Xi^0, (Q - \overline{Q}) = \quad 0 - (-\tfrac{1}{2}) = +\tfrac{1}{2}$$

The two pairs of K mesons can be treated in the same way. The charge, Q, for the η^0 is 0. As η mesons only exist in the neutral form, this is also their 'average value', Q.

$$\therefore \text{For the } \eta^0, (Q - \overline{Q}) = 0 - 0 = 0.$$

The Λ^0 can be treated in the same way as the η^0.

Question 19

Answer

The strangeness is zero and the value of $(Q - \overline{Q})$ is also zero.

Comment

Note that in SU3 groupings of eight particles, there should be *two* particles at the centre of the array. As it stands in Figure 44, the array has only a single member in that location. The missing member has the strangeness and $(Q - \overline{Q})$ values characteristic of that location, i.e. they are both zero. The missing particle is in fact the η^0 meson. At the time Gell-Mann and Ne'eman put forward their SU3 theory, this array *was* incomplete in the way described in this question. The theory was used to predict the existence of the new particle, just as it was used to predict the existence of the omega-minus particle.

Acknowledgements

Grateful acknowledgement is made to the following source for material used in this Unit:

Photo CERN for Figs. 6, 16, 18 and 19; Science Museum, London for Fig. 8.

S.100—SCIENCE FOUNDATION COURSE UNITS